Biofuels and Rural Poverty

Biofuels and Rural Poverty makes an original contribution to the current controversial global debate on biofuels, in particular the consequences that large-scale production of transport fuel substitutes can have on rural areas, principally in developing countries but also in some poor rural areas of developed countries.

Three key concerns are examined with a North–South perspective: ecological issues (related to land use and biodiversity), pro-poor policies (related to food and land security, gender and income generation) and equity of benefits within the global value chain. Can biofuels be pro-poor? Can smallholder farmers be equitably integrated in the biofuels global supply chain? Is the biofuels supply chain detrimental to biodiversity?

Most other books available on biofuels take a technical approach and are aimed at addressing energy security or climate change issues. This title focuses on the socio-economic impact on people's livelihoods, offering a unique perspective on the potential role of biofuels in reducing rural poverty.

Joy Clancy is Associate Professor in Technology and Development at the University of Twente, the Netherlands. She has published extensively on development, energy and poverty.

Biofuels and Rural Poverty

Joy Clancy

LONDON AND NEW YORK

First published 2013
by Routledge
2 Park Square, Milton Park, Abingdon, Oxon, OX14 4RN

Simultaneously published in the USA and Canada
by Routledge
711 Third Avenue, New York, NY 10017

Routledge is an imprint of the Taylor & Francis Group, an informa business

British Library Cataloguing in Publication Data
A catalogue record for this book is available from the British Library

Library of Congress Cataloging-in-Publication Data
Clancy, Joy.
Biofuels and rural poverty / Joy Clancy. -- First edition.
   pages cm
   "Simultaneously published in the USA and Canada"--Title page verso.
   Includes bibliographical references and index.
   1. Biomass energy--Economic aspects. 2. Biomass energy--Social aspects.
   3. Biomass energy--Environmental aspects. 4. Energy policy. 5. Rural
   poor. 6. Poverty--Government policy. 7. Equality. 8. Social justice.
   9. North and south. I. Title.
   HD9502.5.B542C55 2013   333.95'39--dc23
   2012027344

ISBN13: 978–1–84407–719–9 (hbk)
ISBN13: 978–0–203–12847–3 (ebk)

Typeset in Bembo by
Bookcraft Ltd, Stroud, Gloucestershire

MIX
Paper from
responsible sources
FSC   FSC® C004839
www.fsc.org

Printed and bound in Great Britain by
CPI Antony Rowe, Chippenham, Wiltshire

Contents

Abbreviations

BVC	Biofuel Value Chain
CAFTA	Central American Free Trade Agreement
CBO	Community-based Organisation
CDM	Clean Development Mechanism
CO_2	Carbon Dioxide
CONAB	Companhia Nacional de Abastecimento
CSR	Corporate Social Responsibility
DFID	Department for International Development
EPA	Environmental Protection Agency (US)
ESCO	Energy Service Company
EU	European Union
FAO	Food and Agriculture Organization of the United Nations
Forex	Foreign Exchange
GATT	General Agreement on Tariffs and Trade
GDP	Gross Domestic Product
GHG	Greenhouse Gas
GM	Genetically Modified
GMO	Genetically Modified Organisms
GSP	Generalised System of Preferences
ha	Hectare
ICRISAT	International Crops Research Institute for the Semi-Arid Tropics
IDB	Inter-American Development Bank
IEA	International Energy Agency
IFAD	International Fund for Agricultural Development
IFAP	International Federation of Agricultural Producers
IFPRI	International Food Policy Research Institute
IISD	International Institute for Sustainable Development
ILO	International Labour Organization
ILUC	Indirect Land-use Change
IMF	International Monetary Fund
km	Kilometre
km^3	Cubic Kilometre
LPG	Liquefied Petroleum Gas

LUC	Land–use Change
MDGs	Millennium Development Goals
MEA	Millennium Ecosystem Assessment
Mha	Million Hectares
MTBE	Methyl Tertiary Butyl Ether
Mtoe	Million Tonnes of Oil Equivalent
NGO	Non-governmental Organisation
NO_x	Nitrous Oxide
nPAH	Nitrated Polycyclic Aromatic Hydrocarbons
OECD	Organisation for Economic Cooperation and Development
OPUL	Oil Palm Uganda Ltd
PAH	Polycyclic Aromatic Hydrocarbons
PNPB	Brazil National Biodiesel Production Programme
PRS	Poverty Reduction Strategy
RSB	Roundtable on Sustainable Biofuels
RSPO	Roundtable on Sustainable Palm Oil
SO_2	Sulphur Dioxide
TEEB	The Economics of Ecosystems and Biodiversity
UNCTAD	United Nations Conference on Trade and Development
VEC	Village Energy Committee
WFP	World Food Programme
WHO	World Health Organization
WTO	World Trade Organization
WWF	World Wide Fund for Nature

Acknowledgements

I have read many acknowledgements in which the author profusely thanks their publisher for their patience and fortitude in waiting for the manuscript. Now I know why! Tim Hardwick has the patience of a saint. He has suffered long in trying to extract this manuscript out of me which has been subject to various delays including two horse-riding accidents. Tim was too polite to say what he really thought when he was awaiting delivery of the manuscript in January 2012 and I announced that I had broken my arm and couldn't type! I trust that he feels that the wait has been worthwhile.

I would like to thank my colleague Jon Lovett for the contribution he made to this book. First he encouraged me to revisit biofuels after a period of 20 years, which I had not been involved in since the completion of my PhD. The depressing aspect of returning to this field of research was how little seems to have changed for the rural poor, and there has been little learning from the past. On the positive side I have been able to apply what I have learned over the last 20 years to the issue of biofuels and rural poverty, which I believe is to the benefit of the reader. Jon made an early draft of Chapter 4 and has provided me with useful material and valuable comments on other parts of the text. Vicky Marin, currently writing her doctoral dissertation at the University of Twente, wrote an early draft of Chapter 5. Her human rights approach to biofuels and conflict has brought new insights for me. Ada Krooshoop and Barbera van Dalm did a fantastic job in preparing the manuscript including spotting grammatical mistakes (ah the Dutch ability with languages!).

I would also like to thank the Institute of Development Studies (IDS) in Brighton for allowing me to spend a four-month sabbatical period to work on the manuscript. Allister McGregor and his colleagues in the Vulnerability and Poverty Reduction Team made the stay an intellectually stimulating time. While in Brighton it was good to be back in contact again with Andrew Barnett, Dolf te Lintelo and Simon Bachelor. Thanks also to Deborah Shenton who made it a 'fun' time. The sabbatical would not have been possible without the financial assistance from the University of Twente Incentive Fund (*UTstimuleringsfonds*) nor without the support from my colleagues in the Centre for Studies in Technology and Sustainable Development (CSTM). Thank you to Frans Coenen, the head of CSTM, for allowing me the time away from the department and to Jon Lovett and Annemarije Kooijman-van Dijk who had to assume additional duties while I was away.

On a more personal note, thank you to my cousin, Chris Glasman, who was striving to meet a deadline to submit his doctoral thesis at the same time I was trying to complete this book. His understanding of the pressure to finish and his encouragement in getting there were a great help. Finally, and by no means last, special thanks to my partner Giles Stacey who has forgone many a lost weekend and endured delays in starting holidays when I needed 'another few minutes just to finish this paragraph'! I suspect he thinks that completing this book will not be the last time he will hear that phrase.

Joy Clancy

1   Introduction

Biofuels,[1] the liquid fuels derived from plant material (biomass), have long been considered as an attractive substitute for petroleum fuels because they have similar combustion properties and work in existing technology. They can be assimilated into the petroleum fuels supply chain with minimal technical adjustments. Biofuels can be produced, with current commercial technology and feedstocks, on a scale commensurate with the volumes of petrol (gasoline) and diesel for transport. In a number of locations biofuels are already well-established commercial products. Brazil, which has used bioethanol intermittently as part of its fuel supply since the late 1920s, is probably the most well known. In the 1970s there was considerable interest in biofuels in response to the oil crises of 1973 and 1978/9. A few countries[2] began biofuel programmes which had all but disappeared when cheap oil returned in the late 1980s. However, since the turn of the millennium, a number of factors have combined to stimulate interest in biofuel production once more. These drivers can be put into two categories. One category with a strong Northern[3] agenda is linked to energy security, high oil prices and environmental concerns. The second category, with an ostensibly Southern agenda, is one in which actors see biofuel production as a key means of promoting rural development (see for example UNDP, 1995; Kammen et al., 2001). The two agendas are linked and there is some crossover, in the sense that Southern governments are also concerned about high oil prices and Northern governments have to take their powerful farming lobbies into account.

This chapter starts by briefly describing how biofuels are produced and their main uses. It then returns to the issues referred to in the previous paragraph and reviews the arguments for and against biofuels. It finishes by setting the scene for the issues discussed in this book.

Biofuels: a brief introduction

The most well-known biofuels are bioethanol (a petrol substitute) and biodiesel (which substitutes for diesel). Their method of production is from well-established commercial processes and crops using what are called 'first generation technologies and feedstocks' that include sugar crops such as sugarcane, sugar beets and sweet sorghum, and starch crops such as maize, wheat, barley, cassava and sorghum grain for bioethanol; and oilseed crops such as rapeseed, soybeans, palm oil, sunflower and mustard seed for biodiesel. There is increasing interest in what are known as 'second generation feedstocks' based on cellulosic biomass, such as perennial woody plants and grasses. All of the plant growth can be converted into biofuels, whereas for first generation feedstocks only a fraction of the plant material forms a fuel feedstock. Second generation feedstocks are, therefore, considered more efficient since they give a higher yield of biofuel per hectare compared to first generation feedstocks. However, the technology is not as readily available for second generation feedstocks as it is for first. There is now also an interest in 'third generation feedstocks' that are derived from aquatic organisms such as algae. The interest in aquatic organisms is being driven by concerns that there is insufficient land to produce biofuels. However, there is little practical experience with 'farming' such organic material so these feedstocks represent a long-term development.

In this book we will concentrate on first generation biofuels, since these are already well established in the South.

Bioethanol

Overview of production process

Bioethanol is produced as a by-product of a naturally occurring fermentation reaction when yeasts break down sugars. The result is a weak solution of ethanol (5 to 15 per cent by volume). For use as an engine fuel, the solution has to be concentrated to at least 95 per cent if the bioethanol is to be used as the sole fuel. If the bioethanol is to be blended with petrol, to make the mixture known as gasohol, all water must be removed. This concentration process uses the well-established chemical engineering technique of distillation. Indeed, ethanol fermentation and distillation to produce potable alcohol are considered to be among the oldest chemical engineering processes known to man.

The aqueous residue after fermentation and distillation, known as stillage, represents a significant disposal problem both in terms of volume and composition. The volume of stillage[4] produced is 10 to 15 times the volume of bioethanol (Kojima and Johnson, 2005). Direct discharge of the stillage into natural waterways can cause serious environmental damage since the stillage is high in nutrients and can lead to eutrophication. Not only is this an ecological disaster but there are also social-economic consequences. Families who live along waterways and depend on fish and other aquatic plants and animals for food are deprived of important resources as well as a source of commodities. There are alternative uses for the stillage, such as a fertiliser, animal feed or converting it to a combustible gas[5] which can be used to fuel the distillation process. In Brazil, using the stillage as fertiliser has become common practice and there is evidence to suggest that there are productivity increases of up to one tonne of sugar per hectare (Moreira and Goldemberg, 1999).[6] The stillage is high in potassium which reduces the need for expensive fertiliser inputs (Zandbergen, 1993).

Feedstocks

Any plant material containing significant amounts of sugar, or materials that can be converted into sugar such as starch or cellulose, can be used to produce bioethanol. The three main sugar-containing crops which could be grown directly for bioethanol are: sugarcane, sugar beet and sweet sorghum. While the first two are extensively grown commercially as feedstocks for sugar and bioethanol, the latter tends to be a subsistence crop. Sweet sorghum has attracted significant attention, especially when the inclusion of smallholders in production chains is an objective because it is a multipurpose crop, it has high drought tolerance and its cultivation techniques are well known to small-scale farmers particularly in semi-arid areas of Africa where grain sorghum (a relative of sweet sorghum) is extensively grown (ICRISAT, 2007). A number of by-products and agricultural processing residues with a high sugar content, such as molasses, fruit wastes and whey, could also be fermented, which would simultaneously act as a waste treatment process. Indeed molasses is already fermented commercially to produce both potable and bioethanol (Rajagopal and Zilberman, 2007) as are wine residues in the European Union (EU) (CFC, 2007). In 2003, around 60 per cent of bioethanol came from sugarcane, sugar beet and molasses (Kojima and Johnson, 2005).

The sugars in starch crops are in macro-molecular form, as carbohydrates, which are not directly fermentable by yeast. The feedstock, therefore, has to undergo pre-treatment to release the sugars. This adds to the costs of production.[7] The starch sources that are used for producing potable and bioethanol include grains, such as maize (corn), wheat, rice, barley, rye and grain sorghum, cassava (manioc or tapioca) and sweet potato. Other carbohydrate sources that have been considered include potatoes, Jerusalem artichokes and cacti (such as agave).[8] There has been considerable concern expressed about the use of starch crops as feedstocks for bioethanol since they form the staple food for many of the world's poorest people. The issue of food security is discussed briefly later in this chapter and more extensively in Chapter 5.

A technical solution to address the concerns of using food crops as a fuel feedstock is to use cellulose, the most abundant organic material on earth.[9] About 90 per cent of cellulose is found in trees, together with two other macro-molecules: hemicelluloses and lignin. All three compounds can be broken down to yield fermentable sugars. The process is much more complicated than for the other two groups of feedstocks, and at the time of writing it is not yet a commercially mature technology, other than the extraction of cellulose from wood for paper making. However, since the cellulose content of biomass is only between 40 and 50 per cent by mass (Worldwatch Institute,

2007: 46), a large part of the biomass would go unconverted, which would considerably affect the economics of bioethanol production. Therefore, considerable research and development are needed to find low cost chemical or biological methods to render the whole feedstock in a form more amenable to further processing.[10]

A mixture of cellulosic waste material, such as paper, cardboard and fabrics is found in municipal solid waste which makes this a potentially attractive feedstock source. In high income countries of the North, paper makes up around 36 per cent of the weight of municipal solid waste whereas in many countries of the South the percentage is much lower (Twardowska *et al.*, 2004). Agricultural residues are also largely composed of cellulose and to the outsider they can appear a waste waiting to be converted into a valuable resource. However, many residues have important uses for rural communities, for example, in soil fertility and stability conservation, as roofing material and cattle feed. There are some residues, for example, cotton stalk, which are burnt as part of disease control and hence could be considered as non-competitive feedstock. The use of such residues, therefore, needs careful analysis on a site specific basis. There are also ecological concerns raised about the planting of extensive monocultures of trees or perennial herbaceous grasses for cellulose production (Worldwatch Institute, 2007) (see Chapter 4).

Bioethanol as a fuel

The main use for bioethanol is as a transport fuel (see Chapter 6 for other uses). Bioethanol can be used alone or blended with petrol (gasohol) in a slightly modified spark ignition engine. Indeed, Otto, the inventor of the four-stroke spark ignition engine, used ethanol as a fuel in his early engine tests (Clancy, 1991). A litre of ethanol contains approximately 33 per cent less energy than a litre of petrol, meaning that to travel the same distance more litres of bioethanol than litres of petrol are needed. However, ethanol is a cleaner-burning fuel than petrol in terms of particulates, carbon monoxide and carcinogens. Although there can be a slightly higher level produced of nitrogen oxides, the gases linked to acid rain and a precursor of ground-level ozone, better known as smog – a form of urban air pollution which causes respiratory problems (EPA, 1999). Gasohol has better fuel properties than pure petrol.

The percentage of ethanol in gasohol varies from 5 to 25 per cent, with 10 per cent appearing to be the most commonly used (Worldwatch Institute, 2007). In this range, car engines only need slight modification and users in colder climates do not experience starting problems. In Brazil, there are cars which run on pure bioethanol and flex-fuel vehicles which run on any mixture of a gasoline-ethanol blend and hydrous ethanol. Car manufacturers have been able to benefit from Brazil's long experience with using bioethanol and have adapted cars to be able to operate on gasohol.

Biodiesel

Overview of the production process

Biodiesel is derived primarily from plant oils. Biodiesel production involves first crushing the seeds to extract the oils which then undergo a chemical reaction with an alcohol (methanol or ethanol) to yield esters. These esters form biodiesel. The ester from methanol is the most commonly used; not only is it cheaper to produce than the ethyl ester, but it is also easier to separate from the by-products. The prefix 'bio' is slightly misleading if it is meant to imply 'renewable' or 'green' since both the methanol and ethanol are derived from fossil fuels. However, it is reported that Brazil is experimenting with sugarcane-derived ethanol, so that the biodiesel can be considered renewable (Kojima and Johnson, 2005).

The by-product of the ester formation is glycerine which has a commercial value as a precursor for a number of products such as pharmaceuticals, cosmetics, toothpaste and paints. The residue from crushing the seeds can be sold as animal feed.

Feedstocks

A wide variety of oils and fats can be used to make biodiesel. A major distinction is between edible and non-edible oils. Currently, three edible oils dominate as biodiesel feedstocks: rapeseed, sunflower and soybean. In Europe, biodiesel derived from rapeseed oil has dominated, while in the US, it has been soybean oil. At the time of writing, 57 per cent of global oilseed production comes from soybean; however, only a small portion (less than 6 per cent) of the soybean global production is used for biodiesel (Worldwatch Institute, 2007). Soy is grown as a highly mechanised large-scale production system.

The oilseed considered as a biodiesel feedstock tends to reflect a country's commercial agriculture, for example, palm oil in Malaysia and Indonesia. An advantage of palm oil as a feedstock is the high oil yields per hectare compared to other vegetable oil crops (see Table 1.1).

The concerns expressed about using food crops as fuel feedstocks has generated interest in non-edible seeds such as *Jatropha curcas* and *Pongamia pinnata*. These plants are promoted for use in the dry and semi-arid regions of Asia and Africa. Reservations have been expressed, however, about the economic viability of these crops under conditions of low inputs and poor land quality (Rajagopal and Zilberman, 2007). China has been using 40,000 to 60,000 tonnes waste cooking oils annually to make biodiesel, while meat processing plants have been using animal fat residues to run their own vehicle fleets (Worldwatch Institute, 2007).

Table 1.1 Summary of main biofuel-producing countries, production capacity as of 2005/6, future targets and main feedstocks

Country	Biofuel and production capacity as of 2005/6	Future targets – quantity and year	Main sources for biofuel
US	18.4 billion litres of ethanol (2006), 284 million litres biodiesel (2005)	28 billion litres of ethanol by 2012 and 1 billion litres of cellulosic ethanol by 2013	maize and in future cellulosic sources
Brazil	17.5 billion litres (2006)	25% blending of ethanol (has been in effect for long time), 2.4 billion litres of biodiesel by 2013	sugarcane, soybean
EU	3.6 billion litres of biodiesel (2005), 1.6 billion litres of ethanol (2006)	5.75% of transportation fuel on energy basis by 2010	rapeseed, sunflower, wheat, sugar beet and barley
China	1.2 billion litres of ethanol (2006)	na*	maize, cassava, sugarcane
Colombia	400 million litres of ethanol (2006)	10% ethanol blending in cities exceeding 500,000 people since 2006	sugarcane, oil palm
Indonesia	340 million litres of biodiesel (2006)	10% ethanol and 10% biodiesel effective April 2006	oil palm
Malaysia	340 million litres of biodiesel (2006)	5% biodiesel from April 2006	oil palm
Thailand	330 million litres of ethanol (2006)	na*	cassava, sugarcane, molasses
Canada	240 million litres of ethanol (2006)	5% ethanol by 2010 and 2% biodiesel by 2012	maize and wheat
Argentina	204 million litres of ethanol (2006)	5% biofuel by 2010	soybean
India	200 million litres of ethanol	5% ethanol in select cities and 10% biodiesel by 2012**	sugarcane, molasses, jatropha (in future)
Australia	170 million litres of ethanol	350 million litres of biofuel by 2010	wheat and molasses
Japan	insignificant	360 million litres by 2010 and 10% biofuel by 2030	imported ethanol

Source: Adapted from Rajagopal and Zilberman, 2007

Notes
* data not found
** according to Rajagopal and Zilberman (2007), biodiesel policy has not yet passed into law in India and is merely a government preference at this point

Use

The use of biodiesel in diesel engines is not a new phenomenon. Indeed, Dr Diesel, the man credited with inventing the engine which carries his name, is known to have carried out tests on a compression-ignition engine running on peanut oil. He made a prophetic speech in 1912 in which he suggested that diesel from vegetable oils would be as important as diesel from petroleum.[11] Since the middle of the 19th century, there have been various attempts to use biodiesel as a fuel.

At present biodiesel is used as a transport fuel in stationary diesel engines for the generation of shaft power and electricity, and in boilers for heating. There are reports of it being used for lighting and in stoves for cooking, using equipment designed to run on kerosene.[12]

Biodiesel is used as fuel for diesel engines either alone or blended with petro-diesel in the range from 2 to 20 per cent. When used as a blend, there is usually no engine modification required. However, when used alone, modifications are required to take into account the higher viscosity and combustion characteristics of biodiesel compared to its fossil equivalent. Lower temperature operation needs additional attention such as adding a heat exchanger which improves the viscosity of the biodiesel. The nature of these modifications is well within the competence of a good vehicle mechanic (Starbuck and Harper, 2009). A litre of biodiesel contains approximately 5 to 12 per cent less energy than a litre of fossil diesel. The energy content of biodiesel varies because the chemical composition varies with the vegetable oil. While this means slightly higher fuel consumption compared to fossil diesel, there is compensation in better ignition properties giving lower emissions.

The addition of biodiesel to conventional diesel can lead to lower particulate, carbon monoxide and hydrocarbon emissions although there can be an increase in emissions of nitrogen oxides (Worldwatch Institute, 2007). Biodiesel contains no sulphur, which reduces sulphur dioxide emissions. The US Environmental Protection Agency (EPA) considers that biodiesel degrades four times faster than fossil diesel.[13] Diesel spills and leakages are known to be detrimental to aquatic and other living organisms. For example, a 240,000 gallon diesel fuel spill occurred in January 2001 near the Galapagos Islands, off the coast of Ecuador. The fuel spread over an area larger than the city of Los Angeles and killed dozens of animals and birds. It is reported to have taken more than two years to clean up (Lloyd and Cackette, 2001). Fossil diesel is not water soluble. Instead it forms a thin film coating surfaces including those of living things, leading to their suffocation. When fossil diesel leaks from storage tanks, it seeps through the soil and can enter groundwater. Biodiesel is water soluble and hence does not create a surface film like fossil diesel. This property, together with its biodegradability, makes biodiesel especially suitable for marine or farm applications, since it has a lower impact on soil and water in the event of spillages. This is not to say that a large discharge of biodiesel into a body of water will have no effect at all, but the ecosystem would be expected to recover from a shock-loading.

Each ester has its own chemical, and hence fuel, characteristics. The methyl ester has slightly different characteristics from its corresponding ethyl ester. Biodiesel viscosity is sensitive to cold weather, and in low temperature regions, fuel tank heaters and anti-gel agents may be needed. Long-term storage can be problematic with some esters undergoing further chemical change, leading to solidification. This makes the selection of the specific ester important.

It is also possible to use un-modified vegetable oils[14] as a fuel. However, this can require extensive expensive engine modifications such as the replacement of parts containing certain plastics and rubber. (The Worldwatch Institute (2007) quotes a figure of €2,000 or more.) The fuel injection timing may need retarding by 1 or 2 degrees to improve ignition and reduce nitrogen oxides emission. There are commercially available engines,[15] designed specifically for use on un-modified vegetable oils, that pay particular attention to the fuel injection system since this has to respond to an oil with a significantly higher viscosity than fossil diesel.

Selecting biofuel feedstocks

The choice of feedstock is a key decision in determining a biofuels policy since cultivation and harvest of feedstock accounts for 50 to 75 per cent of the total cost of biofuel manufacture, with the highest share for biodiesel production (CFC, 2007). There are also implications for the environment (land, water and chemicals) and social costs (labour requirements, small-scale production versus large-scale agribusiness). Selection of a feedstock may vary within a country to reflect different soil types or rainfall levels. Other countries may already be growing feedstocks which could switch to the biofuel supply chain and are looking for diversification, such as the sugar producers of the Caribbean or tobacco growers in Malawi. The Caribbean sugar industry has recently faced fierce competition from Brazil (a more efficient producer) while at the same time losing its preferential tariff from the EU. Although the region has diversified into financial services and tourism, sugar is still a major employer – for example, 40,000 people in Jamaica; and service providers, such as clinics and social services, in Guyana. There are fears that farmers will be induced into drugs production if no financially attractive alternative is found.[16] Tobacco is Malawi's main cash crop, much of which is grown by smallholders (an estimated 110,000 in 1995–96 (Tobin and Knausenberger, 1998)). The industry has been hit by low prices and farmers are not able to cover the costs of their inputs.[17] A ten-year contract to grow jatropha for biodiesel production was signed in 2004 between a UK company and a tobacco contractor in Malawi. The jatropha is to be grown by smallholders (IIED, 2008).

The water demand of biofuels needs to be part of the assessment. The high water demands of both sugarcane and palm oil are well known (see Tables 1.2 and 1.3), so alternatives which do not rely on irrigation are being promoted for use in drier regions. Sweet sorghum is a multipurpose crop producing grain, sugar (extracted from the stalk) and fodder (from residues). Under certain conditions multiple harvests are also possible (UN–Energy, 2007). Sweet sorghum's production costs have been estimated at 40 per cent less than those of sugarcane (CFC, 2007). Jatropha is promoted as a biodiesel feedstock based on its reputation for low input requirements; for example, it will grow on marginal land in low rainfall areas. However, such conditions will have an impact on yields (Jongschaap *et al.*, 2007) and farmers may well be tempted to increase yields through irrigation.

There are considerable gaps in agricultural knowledge about how many of the plants under consideration as biofuel feedstocks will behave as cultivated crops. Much work needs to be done, therefore, to define what are the optimal inputs, as well as developing new cultivars and the sustainable management of cropping systems together with the long-term environmental impacts for each potential feedstock (Ortiz *et al.*, 2006; Jongschaap *et al.*, 2007). Jatropha is a very good example of where hype[18] does not always match the practice (Jongschaap *et al.*, 2007). Although jatropha is extensively grown in many countries of the Sahel, it is for its live fence properties rather than the oil from the seeds.[19] Animals do not graze the shrub, so the hedge helps combat soil erosion and protects the crops. There is probably some blurring of meaning: 'can survive drought conditions' has been interpreted as 'it grows with little rainfall'. While this is true, the impact on yield is significant, since water uptake by the plant influences flowering and fruit formation. It appears that annual rainfall of at least around 600mm plus fertile soil is needed to provide sufficient seeds to make biodiesel production economic (Christian Aid, 2009). In South Africa, jatropha cultivation does not receive central government assistance because the plant is regarded as invasive, although this is contested.[20] Plants which behave well in one type of ecological system behave as invasive thugs in other ecological systems. This is a good example of the need for research.

Given the significant influence the crop production cost has on the overall fuel cost, efforts at reducing the agricultural component of the production chain can be anticipated. One way of lowering the feedstock cost for a given crop is to increase the yield per hectare. There are a number of possible ways to achieve this, such as: moving to more fertile areas; plant breeding including the use of genetically modified (GM) seeds; and optimal application of fertilisers, pesticides and herbicides. In the South these could be controversial. If moving to more fertile areas displaces food crops, food security could be undermined (see Chapter 5). As Kojima and Johnson (2005) point out, increased use of chemical inputs would be moving counter to the trends in high income countries where the effort is to reduce these inputs and hence costs without lowering yields. GM crops are divisive, although whether the negative reaction from consumers will be as strong for fuels made from GM crops as it is with food remains to be seen. However, GM seeds are grown in the South with Brazil being one of the main users. In 2007, 64 per cent of the country's soybean crop was genetically modified.[21] Mechanisation helped to increase yields in the North but at a price for small-scale producers

Table 1.2 Land and water intensity of potential sources for ethanol

Ethanol feedstock	Global acreage (million ha)*	Water required mm/yr (low)**	Water required mm/ yr (high)**	Crop yield (tonnes per/ha)*	Ethanol conversion efficiency (lt/t)***	Gasoline equivalent ethanol yield (lt/ha)	Ethanol yield per unit of water (lt/mm)	Growing season (months)
Wheat	215	450	650	2.8	340	600	1.09	4–5 months
Maize	145	500	800	4.9	400	450	0.69	4–5 months
Sorghum	45	450	650	1.3	390	450	0.82	4–5 months
Sugarcane	20	1,500	2,500	70	70	3,300	1.65	10–12 months
Sugar beet	5.4	550	750	100	110	7,370	11.34	5–6 months
Sweet sorghum	Insig.	450	650	40	70	1,900	3.45	4–5 months
Bagasse*	na	na	na	18.9	280	3,550	na	na

Source: Rajagopal and Zilberman, 2007

Notes

* estimates that are typically cited, na – data not available or not applicable; insig. – not significant

* data from FAO online statistical database

** data from FAO crop management database http://fao.org/ag/AGL/AGLW/watermanagement/default.stm

*** data from various sources

Table 1.3 Land and water intensity of major oilseed crops

Oilseed crops	Oil content as % of seed wt	Water required mm/yr (low)	Water required mm/yr (high)	Trees per hectare	Average crop yield in kg/ha	Average oil yield in kg/ha	Oil yield per unit of water (kg/mm)	Time to full maturity	Useful life (years)
Coconut	70%	600	1,200	100	na	4,500	5.00	5–10 years	50
Oil palm	80%	1,800	2,500	150	na	5,000	2.33	10–12 years	25
Groundnut	50%	400	500	na	1,015	508	1.13	100–120 days	na
Rapeseed	40%	350	450	na	830	332	0.83	120–150 days	na
Castor	45%	500	650	na	1,100	495	0.86	150–280 days	na
Sunflower	40%	600	750	na	540	216	0.32	100–120 days	na
Soybean	18%	450	700	na	1,105	199	0.35	100–150 days	na
Jatropha★	30%	150	300	2,000	2,000	600	2.67	3–4 years	20
Pongamia★	30%	150	300	3,000	5,000	1,500	6.67	6–8 years	25

Source: Rajagopal and Zilberman, 2007

Note

★ crops not commercially grown, calculations are based on estimates that are typically cited

and landless rural people who participate in labour markets. Mechanisation brings reduction in employment as well as being an unlikely option for small-scale producers (either their land is taken over, or equipment is not of an appropriate scale for their field sizes, or equipment is too expensive). Brazil has shown considerable improvement in agricultural yields through plant breeding and modification, as well as through the use of chemical inputs. In the 1970s, the ethanol yield per hectare from sugarcane was around 2,000 litres and 30 years later this yield has more than tripled (7,000 litres per hectare) (Worldwatch Institute, 2007: 40).

The technology for processing first generation biofuels is mature and there is little in the way of efficiency gains to be found, particularly in biodiesel processing. For bioethanol, there may be energy efficiency gains in both the pre-treatment and distillation stages. The US ethanol yield per kg of maize has been steadily increasing, from less than 2.4 US gallons per bushel in the 1970s to 2.6–2.8 US gallons per bushel by the mid–2000s, depending on maize's starch content and the process efficiency. The utility (energy and water) costs declined from about US$0.06 per litre of ethanol produced in 1970s to US$0.04 per litre in 2000. This was attributed to improved yields and increases in plant size (Kojima and Johnson, 2005). However, there are limits to economies of scale since sugar and oil crops begin to deteriorate rapidly after harvesting. For example, palm oil begins to deteriorate within 24 hours (Vermeulen and Goad, 2006) as does sugarcane[22] which affects the product quality in terms of the chemical composition. Hence this limits the transport distance between the place of growing the crop and the processing plant. Economies of scale do not have to be achieved by increasing the area under production feeding into a central plant, e.g. in Brazil, economies of scale for sugarcane processing have been achieved by increasing the yield per hectare (Goldemberg, 2006). Nor do economies of scale have to be achieved through large estates; small-scale palm oil producers in Malaysia have been shown to have lower production costs than plantations (Vermeulen and Goad, 2006). There are also limits to economies of scale increasing exponentially based on the cost of transporting bulky feedstock materials to a central processing point (Legge, 2008); also there is an assumption in economies of scale calculations that there is sufficient contiguous land to serve single large processing facilities, which might be difficult to achieve in practice.

Brazil is the market leader for bioethanol and has considerably lower production costs than the US and Europe, which in part reflects lower labour costs but also the contribution that the residue after crushing the cane makes to the process energy. Starch crops, such as maize and sugar beet, produce little in the way of residues that can be used to fuel the distillation process and as a consequence have to use fossil fuels (which in itself is controversial – see below). In 2002, bioethanol in Brazil was produced from sugarcane at around US$0.19 per litre whereas in the US it was US$0.23 per litre from maize and in the EU US$0.51 per litre from sugar beet (Kojima and Johnson, 2005). These figures neither represent differences in agricultural practice nor are they representative of the price at the pump. In Brazil not all sugarcane requires irrigation or chemical fertiliser and pesticides, whereas commercial maize and sugar beet use irrigation and chemical fertiliser.

If a biofuel programme is intended to support small-scale producers and other low income rural dwellers, this policy goal affects the economics of production since costs are scale dependent and the costs for small-scale operations are higher than large-scale processing for the same crop. A study by the International Energy Agency (IEA) gives the processing cost of US$0.20 per litre of biodiesel in small-scale plants and US$0.05 per litre in large-scale plants (Kojima and Johnson, 2005). The economics are also sensitive to marketing any by-products. For example, bioethanol from sugarcane uses the residue remaining after crushing the cane (bagasse) to provide the energy for the process (which offsets the need for fossil fuels) as well as selling molasses; while the waste remaining after bioethanol from maize and biodiesel from vegetable oils can be sold after crushing for animal feed or used as fertiliser (FAO, 2008a).

Glycerine, the by-product from refining the vegetable oil to the ester, is also sold as a chemical feedstock and is used in pharmaceuticals, cosmetics, toothpaste, and paints (Kojima and Johnson, 2005). Therefore, who owns the processing part of the production chain, and hence reaps the rewards, is also an important consideration if the objective is to benefit the rural poor. However, over-production can have a negative effect by flooding the market and reducing the by-product price (as has happened with glycerine from plant oil processing).

Some crops are traditionally grown on extensive plantations, e.g. sugarcane.[23] Feedstocks that are considered particularly appropriate for smallholder production include sweet sorghum and jatropha. An advantage of sweet sorghum is that its cultivation is similar to grain sorghum, which is a staple crop in many parts of Africa, hence small-scale farmers might be more willing to adopt the crop rather than a species that is completely new to them and requires new farming techniques and extra inputs. Jatropha has to be hand harvested, as do many of the oilseeds from trees, and this is an advantage when increasing rural employment opportunities. Employment creation from agricultural production can be used as a feedstock selection criterion.

There has been a debate since the 1970s, as to whether or not biofuels use more energy in their production than is contained in the fuel produced. All fuels, including fossil fuels, use some energy all along the production chain. However, the ratio of energy out to energy in (known as the energy balance) should be positive to be considered sustainable. The methodology for calculating this ratio is complex and can cause as much argument as interpretation of the value calculated! Values vary with the feedstock, also the methods of cultivation and processing. A particular feedstock can also have different values depending on the location where it is produced. For example, bioethanol from sugarcane in Brazil has an energy balance varying between 3.7 and 10.2 units (Dufey, 2006). The inclusion or exclusion of by-products, as well as the assigned value for energy, also influences the value calculated for the energy balance. A study in the US of bioethanol from maize and biodiesel from soy found that maize has a positive energy balance of 1.25 when the energy content of the residue (distillers dry grain with solubles – although this product is usually sold for animal feed rather than combusted) is included (Hill *et al.*, 2006). There are differences of opinion where to draw the boundaries of the analysis; for example, should the energy required to produce the tractor used in land preparation be included?

A large part of the literature on the energy balances has been focused on maize produced in the US. A lengthy debate has been taking place between Professor David Pimentel from Cornell University, New York, and different US Government agencies about whether or not bioethanol from maize yields more energy than the energy contained in the inputs to produce it (see for example Shapouri *et al.*, 2002; Pimental, 2003; Wang, 2005; Pimental and Patzek, 2005). This debate is possibly irreconcilable since it is between different discourses. Those who oppose bioethanol from maize tend to see it as a moral issue: the use of a food crop, in a hungry world, to make fuel. However, the supporters of biofuels tend to see it as an economic issue in the need to stimulate the rural economy.

Ideally, the energy balance can *help* in selecting the crops with the highest yields. However, we doubt if this methodology is used in practice by those making decisions about biofuel feedstocks. Neither UN-Energy nor the Common Fund for Commodities (CFC) in their Frameworks for Decision Makers refer to energy balances as part of selection criteria (CFC, 2007; UN-Energy, 2007). This may be due to the complexity of the methodology, including the need to make calculations on a case-by-case basis and the fact that for a number of feedstock plants there is no data and the data that do exist are controversial (see for example Pimental, 2003). The methodology has possibly been overshadowed by a different argument that is currently higher on the global agenda: reducing greenhouse gas (GHG) emissions (the links between greenhouse gases (GHGs) and biofuels are discussed below). Yet, the energy balance is only one issue at stake in the decision about whether or not to produce biofuels. We would argue that the decision whether or not to grow biofuels should be based on more than one criterion.

Trends in biofuel production

In 2006, global biofuel production was estimated at over 55 billion litres of which 93 per cent was bioethanol and 7 per cent biodiesel (CFC, 2007). Countries from the South are amongst the major producers. Brazil was responsible for 41.1 per cent of global ethanol production and was the second leading producing nation, while Malaysia was the seventh leading producer of biodiesel with 2.2 per cent of the market (Worldwatch Institute, 2007). However, this is a drop in the ocean compared to the annual global production of petrol (approximately 1,200 billion litres (Dufey, 2006)).

There is a distinct global division where biofuels are used and produced. Biodiesel has been predominantly used in Europe and bioethanol in the Americas. An overview of the major biofuel-producing countries and their main choice of feedstock has been given in Table 1.1. If the EU is taken as a single market, it can be considered that most of the biofuels produced are for domestic consumption with a small portion going for export. In the case of bioethanol only about 10 per cent of global production is exported (Dufey, 2006), while the figure for biodiesel is uncertain because the trade statistics are, at least at the time of writing, not accurate (CFC, 2007). However, in the not too distant future, bioethanol production is expected to change to serving export markets. This is partly driven by trade agreements between the major biofuel-using areas and producers, for example, the Central American Free Trade Agreement (CAFTA) which permits bioethanol imports from eligible countries[24] into the US. A number of the countries act as a re-processing facility for Brazilian ethanol, enabling Brazil to benefit from favourable terms from which it is normally excluded. Another driving force is increasing demand from countries with low production potential, such as the EU, Japan, Korea and Taiwan, and an insatiable demand for liquid transport fuels, the US. It is impossible to give an exact prediction about the quantity of biofuels likely to be produced in the future, because there are so many influencing factors, such as energy policies, decisions by car manufacturers about alternative fuels (e.g. hydrogen or electric vehicles), and public attitudes. However, the IEA predicted that world biofuel production would quadruple to over 120,000 million litres between 2005 and 2020, accounting for about 6 per cent of world motor petrol use, increasing to 10 per cent by 2025 (IEA, 2004).

Brazil has long been considered the market leader in bioethanol production with commercial fuel production dating back to the 1920s, although serious investment in bioethanol as a fuel began in the 1970s with the launch of the PROALCOOL programme.[25] Production is mainly orientated towards the domestic market (less than 15 per cent of the 15 billion litres annual production goes for export (CFC, 2007)). Such is Brazil's experience in bioethanol technology that it has positioned itself as a leading exporter of processing technology. However, in 2006, Brazil lost its position as the global leader of bioethanol production when it was overtaken by the US. The US has been producing bioethanol since the 1970s. The output has fluctuated in line with federal government support to maize farmers. Production increased from 4 billion litres in 1996 to around 20 billion litres in 2006 (Dufey, 2006). A number of countries where sugarcane is a traditional crop, such as in the Caribbean islands, Central and Southern America (Costa Rica, Nicaragua, Guatemala, Colombia, Peru and Paraguay) Africa (South Africa, Malawi, Zimbabwe, Mozambique, Ghana and Kenya) and Fiji are investing in bioethanol production. In Europe, bioethanol is produced primarily from sugar beet accounting for 7 per cent of global production (CFC, 2007). In Africa and India, due to the high water demand of sugarcane, there is interest in sweet sorghum which has a much higher drought tolerance. A small number of countries use cassava (China, India and Thailand) and grains such as rice (China) and barley (France).

The commercial production of biodiesel began in the 1990s, although the technology has been available for much longer. Since 2000, the production of biodiesel has been growing at an average annual growth rate of 33 per cent reaching 3.7 billion litres in 2006 (CFC, 2007). The EU is the main centre of production, in 2005 accounting for 90 per cent of global production, primarily from rapeseed. However, the price of rapeseed oil increased dramatically reaching just over US$1,700 per metric tonne in July 2007.[26] This has created a window of opportunity for other oilseeds, most notably palm oil. While Malaysia and Indonesia are the world's leading producers of palm oil, other tropical countries, such as Thailand, Brazil, Colombia, Ecuador and some of the Pacific Islands are showing interest in the palm. Coconut oil is the biodiesel feedstock planned in Philippines and some of the Pacific Islands while the US and Brazil use mainly soybean.

India has chosen to focus on a number of non-edible oils, with jatropha the most extensively grown. The low water and fertiliser inputs of jatropha have attracted the attention of a number of African countries such as Burkina Faso, Mali, Ghana, Malawi and South Africa.

Making comprehensive predictions about the future of biofuels as an export commodity is difficult due to the complexity of the production chain, which starts in the agriculture sector and finishes in the energy sector with an extensive array of semi-processed products in between. There is also the possibility with first generation biofuel crops to switch from the fuel chain back to supplying the agricultural chain. The decision to export as part of the fuel or agriculture supply chain will be influenced by a number of factors including the incentives provided by importing countries. Depending on how it is classified (agricultural product or fuel or chemical), a commodity will fall under different international trade rules, which contributes to uncertainty. Indeed, the members of the World Trade Organization (WTO), which governs international trade, cannot agree whether biofuels are agricultural or industrial goods (UNCTAD, 2008). There are also a plethora of bilateral trade agreements which offer different tariffs. These tariffs can also influence whether or not the commodity is exported as raw or processed material. For example, both the US and the EU have no tariffs for the import of soybeans, while (at the time of writing) they charged tariffs of 19.1 per cent and 8.8 per cent respectively for soya oil (Dufey, 2006: 26). The implications of which are that it is more favourable to a feedstock-producing country to export the seeds rather than refine the biodiesel, while the value added from refining will accrue to the importing countries. Tariffs can vary in response to domestic political pressure in importing countries meaning that the exporting country might have made an expensive investment mistake. For example, Pakistan had risen to be the second largest ethanol exporter to the EU, benefiting from a zero import tariff under a Generalised System of Preferences (GSP) intended to combat drug production and trafficking. Suddenly a 15 per cent import duty was imposed since ethanol no longer qualified under that particular GSP (Doornbosch and Steenblik, 2007). As a result two of the seven operating distilleries were closed, and another five new distilleries were put on hold due to uncertain market conditions (Dufey, 2006). The change in Pakistan's status was attributed to political pressure from ethanol producers within the EU as well as from India who complained to the WTO about the way the GSP operated (Bendz, 2005).

Other uncertainties are created by the development of the so-called second and third generation biofuel technologies which would pass the competitive advantage to the North where control of the technology would lie. In 2005, with the issuing of its Biomass Action Plan, the EU appeared to be shifting its interest from first generation biofuels to second and third generation (Dufey, 2006). The main export market for biofuels is as transport fuel in the North. However, other alternative fuels, such as hydrogen and electricity, are also under consideration, which could be readily produced in the North.[27]

However, these uncertainties do not seem to be deterring countries in the South from implementing or considering biofuel programmes, or in Brazil's case expanding its existing programme (see for example the Action Plan for Biofuels Development in Africa (IISD, 2007)); for bioethanol – Costa Rica, Guatemala, Nicaragua, Thailand, China, Kenya, Malawi, Zimbabwe and South Africa; for biodiesel – Malaysia, Indonesia, Thailand, Philippines, Burkina Faso, Cameroon, Ghana, Lesotho, Madagascar, Malawi, South Africa, Swaziland, Zambia and a number of the Pacific Island states (CFC, 2007). At the end of 2010, 39 countries were preparing or had already implemented mandatory biofuels programmes (Hart Energy Consulting and CABI, 2010). For a number of the sugar-exporting countries already having large areas of land under cultivation, it offers an opportunity to diversify markets. The reform of the EU preferential access for a number of sugar-producing countries has very serious consequences; in the case of the Caribbean this could mean a loss of up to 40 per cent of sugar revenues (CFC, 2007). It is quite possible that feedstock-producing countries will switch to supplying internal markets, which is already the primary target of Brazil's ethanol and India's biodiesel programmes, or develop South–South trade, for example, Philippines, Indonesia and Thailand see India, China and South Korea as large growing markets. There are also studies in Africa taking place about the potential for regional biofuels trade (Takavarasha *et al.*, 2005; UNF, 2008).

Certainly there are a number of indications that in the short to medium term biofuel production will increase (see for example Worldwatch Institute, 2007; IEA, 2009; FAO, 2008a). The IEA, probably one of the most often quoted predictions of biofuel trends, has forecast that by 2030 biofuels will provide between 102 million tonnes of oil equivalent (Mtoe) and 164 million Mtoe of transport fuels. This level of output will require between 34.5 and 52.8 million hectares (Mha) of land (FAO, 2008a). In the next section, we look in more detail at the factors that are driving this process and the concerns that the increase in biofuels use brings.

Biofuels: to grow or not to grow

Governments are keen to promote biofuel production as one of their energy options. This has mainly taken the form of setting targets for blending with petroleum–based fuels, supported by financial incentives to different parts of the supply chain. For example, the EU has set a target of biofuels providing 5.75 per cent of transport fuels by 2010, which it has supplemented with a Strategy for Biofuels that sets out the measures to help reach that target (Kutas *et al.*, 2007). The European Council in 2007 set a specific target for biofuels at a 10 per cent binding minimum to be achieved by all member states by 2020 (Edwards *et al.*, 2008). Meanwhile, the US aims to have nearly 30 billion litres of biofuels being used for transport fuels by 2012.

Brazil is not the only Southern country to produce for its internal market. For example, Colombia introduced a blending mandate of 5 per cent biodiesel in cities over 500,000 (Dufey, 2006) and even oil-rich Venezuela will produce bioethanol for a 10 per cent blend with petrol (Worldwatch Institute, 2007). There are similar stories in Asia. For example, Thailand aims to reduce the cost of oil imports with a 10 per cent blend of bioethanol in petrol, and has a number of policy goals driving the interest in biofuels.

This next section reviews the debate taking place about the merits of arguments used to justify biofuel production and use. First it explores the policy goals from Northern and Southern perspectives. This is followed by an examination of concerns about the impacts that the rapid expansion of biofuels is having, primarily in the South. Here we only present the proponents' perspectives. A number of their arguments are challenged but we will return to these in more detail in later chapters where we examine the veracity of the claims by the biofuels pro- and anti-lobbies from the perspective of the impact on rural poverty in the South.

The Northern agenda

Oil prices

High oil prices and energy security are issues for all governments since they undermine economies that have become increasingly dependent on oil, especially for transport. For oil-importing countries, ensuring an uninterrupted supply at as low a price as possible are key economic objectives, which in turn becomes linked to political objectives (see next section).

The first decade of the 21st century has seen a dramatic rise in oil prices. From US$16/barrel for crude oil in January 1999, the international spot market price began to rise to previously unknown highs with a peak of US$145/barrel in July 2008.[28] A number of factors have contributed to this price rise. The demand for oil has increased (5.3 per cent between 2002 and 2004) and this demand now exceeds the global installed refinery capacity, which would take time to increase, even if there were the desire to do so (Worldwatch Institute, 2007). While it is not unusual to blame emerging economies – India and China in particular – for the increase in demand, a number of industrialised countries have also shown an increase in demand (e.g. 6.3 per cent in the UK and 10.2 per cent in Canada between 2002 and 2004) despite government commitments to reduce oil consumption.[29] Another factor in the oil prices increase is the vulnerability of oil extraction to the vagaries of the weather, such as hurricanes in the Gulf of Mexico,[30] and political instability linked with civil unrest, such as the so-called Arab Spring, disrupting supplies leading to price increases.[31]

The price of oil has a significant influence on the performance of an economy. For oil-importing countries, it has an impact on the trade balance. Oil has to be paid for in dollars which have to be earned from exports. This requirement can put considerable strain on developing countries' economies that have weak currencies. Of the world's 50 poorest countries, 38 are net importers of oil and 25 import all of their petroleum requirements (UN-Energy, 2007: 39). The situation is often most dire for landlocked countries, which include many of the world's poorest countries – for example, Malawi, Zambia and Nepal – since they pay more than the international spot market price because of the additional transport costs (up to 50 per cent). An indication of the impacts on developing countries can be seen from the International Monetary Fund's (IMF) estimate of impact on the Indian economy. The Fund estimates that for every US$10 per barrel [sic] increase in the price of oil, the gross domestic product (GDP) will decrease by 1 per cent (TERI, 2004). Potentially, biofuels can represent considerable savings on foreign exchange (forex). It is estimated that the bioethanol programme in Brazil has saved €50.2 billion on oil imports between 1976 and 2004 (Worldwatch Institute, 2007). Higher oil prices, which are linked to higher food prices, also contribute to inflation (see below).

For oil-exporting countries, a use has to be found for all the extra income. This can take the form of legitimate investment in financial markets or infrastructure at home. The underside is political de-stabilisation and corruption, as has been seen in a number of countries such as Angola, Ivory Coast and Sudan. Very large amounts of income suddenly become available from one sector (in this case oil) that distorts the economy leading to contractions or lack of development in other sectors, with negative effects on income and jobs. In Nigeria, for example, the oil boom of the early 1970s caused agricultural exports to drop from 11.2 per cent of GDP in 1968 to 2.8 per cent of GDP in 1972 (Ross, 2008). When the oil is located in one region, this can lead to calls for greater autonomy from the central government by the oil-rich region which can develop into armed conflict funded by oil revenues (e.g. Angola where the protagonists were able to buy military jets (Frynas and Wood, 2001)) or by extortion (e.g. Colombia where oil companies became targets for guerrilla groups (Pearce, 2007)) or if the benefits are not felt in other regions (e.g. Sudan where discovery of oil in the south led to conflict over where control of revenues should rest (Ylonen, 2005; Ross, 2008)).

Biofuels are seen as a renewable alternative to fossil fuels. Not only are the fuel characteristics similar so they can be used in existing engines with little modification (see above) but biofuels can also be integrated into existing fossil fuel distribution systems with minimum disruption. Other alternative transport fuels, such as hydrogen and electricity, require more radical alterations to fuel distribution systems which would be expensive to implement.

Energy security

Northern governments have the memory of an elephant when it comes to the economic and political effects of the oil embargos of the 1970s and the need to ensure securities of supply. In 2000, oil imports of Organization of Economic Cooperation and Development (OECD) countries accounted for 52 per cent of their energy requirements, but this is expected to rise to 76 per cent by 2020 (Dufey, 2006). Northern governments do not sleep any easier in their beds knowing that approximately 80 per cent of oil reserves are controlled by state-owned oil companies located in a small number of countries generally having weak or non-existent democracies with their own political agendas in respect of Northern governments, in particular the United States (US). The need for liquid fuels to keep their economies operating, leads Northern governments to look for other options. This can be as extreme as military intervention but there are also technical solutions. Improving energy efficiency is one option, while the high price of oil makes the expensive options of oil shale and coal to oil conversion attractive. However, the latter brings strong opposition from environmentalists.

Biofuels offer an opportunity for diversification within the energy supply mix and improve energy security and reduce vulnerability to oil price shocks. The effects of oil price rises on an economy are complex and a number of effects can occur, as was seen in the 1970s after the two oil crises (Cloin *et al.*, 2007). Inflation arises when the oil price rise is passed on to consumers, which leads to demands for wage increases adding to inflationary pressures. Demand for goods and services may decline with increased prices, which would be damaging for the economy if they are locally produced; although imports might help with the balance of payments. If the exchange rate remains unchanged the balance of payments will be negatively affected since more export income has to be earned to pay for the same volume of oil imports. If petroleum products are subsidised, which is not uncommon, then this can put pressure on the government's financial resources if subsidies are increased. Low income households tend to spend a larger part of their income on purchased fuels than higher income households. Therefore, in countries where petroleum fuels are subsidised, not matching oil price rises with subsidy increases could potentially be damaging for low income households.

Energy security can be considered from the local level perspective as well as the national level. Biofuels can be produced at a price that is competitive with petroleum fuels particularly where geographical isolation adds to transport costs for the latter, for example, Caribbean and Pacific Islands and landlocked countries (Cloin *et al.*, 2007). Access to a reliable supply of affordable modern energy sources can be particularly significant for the rural poor (see Chapter 6).

Climate change

Northern governments are concerned about how to fulfil their commitments to reduce GHG emissions as part of the measures to combat climate change. Transport consumes around one-third of global energy and fuels and accounts for about 21 per cent of annual GHGs (Rajagopal and Zilberman, 2007). Biofuels offer the option of displacing fossil fuels, particularly for transport where the share of emissions continue to grow. Indeed, in the EU, growth in emissions from transport is considered to be the main reason for the EU not meeting its Kyoto commitments (Biofuels Research Advisory Council, 2006).

The EU approved two draft directives in 2003 concerning energy supply diversification and the reduction of GHG emissions. Indicative targets for member states have been set for biofuel consumption in the transport sector: biofuels must constitute 2 per cent of all petrol and diesel motor fuels by 2005 and 5.75 per cent by 2010 (Dufey, 2006).

The basic argument is that because growing feedstocks absorb carbon dioxide (CO_2), the release of CO_2 emitted during biofuel combustion does not contribute to new carbon emissions since the emissions are already part of the fixed carbon cycle. However, there is considerable variation in GHG savings – ranging from negative to more than 100 per cent. Estimates vary according to the type of feedstock, cultivation methods, conversion technologies, energy efficiency assumptions and disparities regarding reductions associated with co-products (Doornbosch and Steenblik, 2007). An evaluation of six studies on GHG reduction of corn-based bioethanol found a variation from a 33 per cent decrease to a 20 per cent increase, averaging a 13 per cent reduction in GHG emissions compared to petrol (Koonin, 2006). Decision making about selection of feedstocks is not helped by the lack of clarity about values to use in making assessments of GHG reduction. It has also been suggested that total amounts of GHG saved might not represent the best assessment approach but the GHG saved per hectare of land, since land availability can be a constraining parameter (Edwards *et al.*, 2008)

Climate change initiatives open up the opportunity for the South to benefit. Germany is purchasing carbon credits from Brazil as part of its Kyoto Protocol commitments and, in turn, helps Brazil subsidise taxi drivers and car hire companies by 1,000 reals (US$3,000) per vehicle on the first 100,000 vehicles sold (Dufey, 2006).

Air quality

When refined petroleum fuels are combusted a number of chemicals are emitted that are considered harmful to health. There are gaseous pollutants, such as sulphur dioxide (SO_2), nitrogen oxides (NO_x) and carbon monoxide, that are linked to respiratory effects, as well as a group of chemicals that are either known human carcinogens, for example, benzene, and probable human carcinogens, such as formaldehyde, acetaldehyde, and 1,3-butadiene

(USEPA, 1994). Particulates[32] in the air are linked to heart and lung disease and those from diesel are suspected carcinogens (Cohen *et al.*, 2004). The primary sources of emissions are vehicles and power stations. While these pollutants are usually associated with outdoor urban air, households using petroleum fuels, such as kerosene for cooking and lighting, will also be exposed to chemicals with potential health impacts (Dasgupta, 1999; ESMAP, 1999). The nature of the health effects depends on a wide range of factors, including length of exposure time and any underlying health conditions. Children are more vulnerable than adults. The urban poor in the South suffer disproportionately from the impacts of air pollution because they tend to live in 'pollution hotspots', that is, in areas with higher concentrations of roads and industrial areas, which are residential areas higher income groups can avoid.

One of the driving forces in the US for biofuels has been environmental legislation linked to urban air quality. The United States Environment Protection Agency (USEPA) considers that engines running solely on biofuels or on a blend of standard fuels and biofuels tend to have lower emissions of particulates, carbon monoxide and sulphur oxides (Dufey, 2006). Biodiesel emissions, compared to petroleum diesel, show decreased levels of polycyclic aromatic hydrocarbons (PAH) and nitrated polycyclic aromatic hydrocarbons (nPAH) (75 to 85 per cent with the exception of benzo[a]anthracene, which was around 50 per cent). PAH have been identified as potential cancer causing compounds.

Ethanol is an ideal substitute for lead which acts as an anti-knock agent in petrol. The use of lead in petrol has declined, and in many countries has been banned as an additive, as a result of the connection between airborne lead and serious health conditions linked to damaged nerve systems, particularly in children (Worldwatch Institute, 2007). Initially, methyl tertiary butyl ether (MTBE) was preferred to ethanol as an additive since it is cheaper to produce. When blended with petrol, MTBE and ethanol both help to reduce photochemical smog, which can be problematic under certain geographical and weather conditions. However, MTBE was found to be a carcinogen and there was a concerted campaign in the US to have it banned. MTBE's high solubility in water raises concerns about it entering waterways.[33] MTBE was banned in the US in the early 1990s, which has provided a stimulus to the bioethanol market.

The Southern agenda

Rural development

The global production of biofuels doubled between 2002 and 2007 and is expected to double again by 2011 (UN-Energy, 2007). As a consequence, some have seen the demand for alternative fuels in Northern markets as an opportunity for product diversification in tropical agriculture, providing an entry into new end-markets. Tropical crops generally have better returns in terms of biofuel yield per hectare than temperate crops[34] and with generally lower production costs, the South could have a competitive advantage over the North. Such opportunities provide a stimulus to rural development. This point is not lost on Northern governments. The US's biofuels programme is in part a response to lobbying by the agriculture sector.[35] The EU has also promoted biodiesel as an instrument to boost rural areas (European Commission, 2006).

Diversification of agricultural commodities is seen as a means of stimulating rapid economic growth (World Bank, 2007). Biofuels fit the profile in terms of diversification for biomass beyond the traditional uses of food, feed and fibre. Such diversification could also help reduce the volatility of commodity prices (CFC, 2007). They also offer an alternative outlet for countries facing problems with changes in trading regimes (sugar producers from the Caribbean and Africa), changes in consumer behaviour (tobacco) and drugs related crime (e.g. Colombia). The export market looks potentially large since Northern countries are unlikely to be able to meet their domestic demand. For example, the EU's goal of 5.75 per cent biofuel content in the fuel transport blend by 2010 will require 18.6 Mtoe of biofuels which cannot be met solely by production internally within the Union (Dufey, 2006). There are also attractive possibilities in the growing markets of India and China. For example, in China the present rate of expansion of biofuel production capacity is only about half of what projections show are required to meet a target of 10 per cent biofuels blend in all Chinese cars by 2020 (22.7 metric tonnes of biofuels) (Dufey, 2006).

Poverty

Seventy-five per cent of the world's poor rely on agriculture for their living (FAO, 2008a). Agricultural economists have long considered that agricultural growth has a significant impact on reducing rural poverty and is more effective in this respect than other sectors of the economy (FAO, 2008a). Based on experiences in Latin America and South Asia, this policy thinking is and was one of the arguments the Commission for Africa used to recommend diversification of non-traditional agricultural export crops to promote growth for poverty reduction, citing Kenya, Uganda and Ethiopia as good role models (Commission for Africa, 2005: 225). As was mentioned in the previous section, biofuels provide a diversification from traditional agricultural export crops. The link between growing biofuels and poverty reduction certainly found favour with a number of African energy ministers at a high-level meeting in Addis Abba in 2007 held to discuss the possibilities for developing biofuels on the continent (IISD, 2007).

The argument for biofuel production to benefit the poor is based on the assumed stimulus that an indigenously produced product contributes to the national economy not only through product diversification (in the case of biofuels) but also the forex benefits (see above). The latter can accrue from direct substitution within the national fuel supply chain (hence reducing the oil import bill and saving forex) but also the potential for export earnings (hence generating forex). It is estimated that Brazil, between 1975 and

2006, saved more than US$65 billion on oil imports (Moreira, 2006). The Colombian Government estimates a 3 per cent increase in GDP bioethanol production from sugarcane.[36] This benefit can be contrasted with the impact of oil price rises on poverty. The 2005 hike in oil prices is estimated as having reduced the GDP growth of net importing countries from 6.4 per cent to 3.7 per cent. As a consequence, the number of people in poverty is estimated to have increased between 4 and 6 per cent (CFC, 2007).

Much of the biofuel system's added value can be generated by processing the feedstock close to the source of production and hence income generation is retained locally (Kartha *et al.*, 2005). Therefore, so the argument runs, biofuels can contribute to rural poverty reduction through job creation and increased incomes through participation in the biofuels supply chain. Smallholders could potentially benefit from growing biofuels feedstocks. There would also be increased employment opportunities in processing and transportation. Large-estate grown crops could offer employment opportunities for unskilled agricultural labour and landless people.

Biofuel production generally requires intensive labour, resource and land inputs. As a consequence they have a direct impact on the communities where they are embedded. These impacts will be on labour demand, land ownership and land use, all of which have strong poverty links. Therefore, adjustment of these parameters can either result in people moving in or out of poverty or deepening their poverty. In terms of land use it is not only the nature of the crops produced but also induced land-use changes from biofuel programmes, with any attendant environmental impacts needing to be assessed (see Chapter 4).

Job creation in rural areas has been one of the major objectives within the European Biofuels Programme. One estimate gave the number of jobs to be created as between 45,000 and 75,000 if biofuels provided 1 per cent of the fuel supply in Europe (Worldwatch Institute, 2007). In the US, it is claimed that biofuel production is now responsible for creating more employment in rural areas than any other activity (Dufey, 2006). Again in Brazil, rural employment has been a major objective of the biofuels programme and there are claims of large numbers of jobs being created, up to 1 million, although it is not clear how many are actually new jobs (Kartha and Larson, 2000; Moreira, 2006). The type of feedstock also influences the level of job creation. Oilseeds tend to be most beneficial in terms of the numbers of jobs created since they are the least mechanised of the biofuel feedstocks.

Increased incomes bring an indirect stimulus to the local economy by creating a demand for goods and services. This is a strong argument for keeping as large a portion of the processing as possible in rural areas. The biofuels produced can also contribute to improving rural energy supplies, which can directly and indirectly contribute to reducing poverty. This is an objective of the Indian biofuels programme (TERI, 2004).

Biofuel feedstocks which allow intercropping of vegetables can help improve rural households' well-being, through improving their food security as well as income from the sale of any surplus (see Chapter 6).

The opposition

However, biofuels have not been universally welcome. There have been a number of concerns expressed about the negative aspects of biofuels.

Food versus fuel

The food versus fuel is not a new debate and was already a major issue when biofuels were being promoted after the oil price rises of the 1970s.[37] More recently, the Special Rapporteur to the UN on the Right to Food is reported as saying that 'the use of agriculturally productive soil for energy crops [is] a *crime against humanity*'.[38] Unfortunately, this quote has tended to be used selectively and appears as 'biofuels are a crime against humanity'[39] which is not quite the same thing. Nevertheless, the food versus fuel issue is probably the most widely voiced concern in relation to biofuels. In recognition of these concerns we devote a whole chapter to the issues (Chapter 5) and so here we only summarise the main arguments.

There is a concern that biofuels compete with and displace food crops from high grade agricultural land, leading to reduced food availability and increased food prices, which affects food access particularly for the poor (UN-Energy, 2007). An additional concern is related to the water demand of biofuels which could put the feedstock crops in competition with households, livestock and food crops for water supplies. A reduction in clean water for households can have negative health implications for family members (UN-Energy, 2007). Ill health increases household vulnerability and decreases capacity for self-sufficiency (see Chapter 2).

The statistics that have caused a lot of unrest relate to commodity prices and the linkages that have been made with the increased production of biofuels. Real food prices in 2008, according to the Food and Agricultural Organization (FAO), were 64 per cent above the levels of 2002 (FAO, 2008a). There was a significant surge of 97 per cent in vegetable oil prices during that period, which has been linked to the demand for biodiesel.

Food crops being diverted to biofuel feedstocks has been linked to increased commodity prices, especially maize. The US is a major maize exporter, particularly to a number of Central American countries, such as Guatemala, Costa Rica, El Salvador and Honduras, where maize is a staple food. In Guatemala, maize provides an average 40 per cent of an individual's calorific intake (Christian Aid, 2009). The price of yellow maize in the US increased from US$1.98 per bushel in January 2006 to US$3.90 per bushel in March 2008. Such near doubling of the price can have devastating effects on poor people who already struggle to meet their daily nutritional intake. There are estimates that suggest that every percentage increase in real prices of staple foods results in 16 million more chronically hungry people, which would bring the total in 2025 to 1.2 billion people (Raswant *et al.*, 2008).

First generation biofuel crops will compete with, and may displace, food crops since they both need good quality land to produce good yields. OECD/

FAO estimated that in 2004 about 0.9 per cent of global cropland (13.8Mha) had been used for crops for transport biofuels (OECD/FAO, 2007). By 2008, the area dedicated to these crops had increased to about 35.7Mha or 2.3 per cent of total cropland most of which was in the US and the EU (UNEP, 2009). There is some uncertainty about exact amounts of land used related to the way in which the areas are estimated.[40] There are signs that in some countries biofuel crops have begun to displace food crops. In 2008, the area of sugarcane in Brazil increased by 9Mha, expanding into grassland as well as displacing crops (UNEP, 2009). Similar impacts from soy have also been reported. However, this illustrates another difficulty when trying to make sensible statements about the impact of biofuels. A number of crops can switch easily between food and fuel chains to take advantage of favourable prices. Indeed many Brazilian sugar processing plants are constructed to do just that – molasses can be switched between animal feed and ethanol fermentation depending on current market prices.

It is a moot point as to whether or not the growing of biofuels in the EU has displaced food crops because farmers are encouraged to grow biofuel crops on land which has previously been taken out of food production (known as set-aside land). The rapid expansion of palm oil in South East Asia has not primarily been at the expense of crop land but of virgin forest, as the latter is regarded as needing less fertiliser and hence is more profitable than cultivated land (UNEP, 2009).

The predictions about how much land would be needed in the future for a world dependent on biofuel transport fuels varies considerably depending on the assumptions made in the models designed to predict likely demand and yields. The IEA estimated that by 2030 between 35Mha of land (2.5 per cent of available arable land) and 53Mha (3.8 per cent of available arable land) would be required whereas other researchers estimated that from 2050 onwards 1,500Mha, equivalent to the current total global farmland, would be required (IEA, 2006; Field *et al.*, 2007). Some countries, such as India, have decided that the use of agricultural land is unacceptable for biofuel production and have identified land considered 'degraded or marginal'. This classification of land as 'degraded or marginal' is contested by rural communities for whom natural ecosystems form an important source of goods and services for rural households (Kartha and Larson, 2006) (see Chapter 4).

Environmental impacts

It is feared that a rapid expansion in growing biofuels could bring both environmental impacts (such as water use and pollution) and social impacts (such as land rights conflicts) (Raswant *et al.*, 2008). The 'adverse effects on biodiversity and human well-being' of biofuels have been identified as likely to occur when their production is associated with:

1   The loss, fragmentation and degradation of valuable habitats such as natural and semi-natural forests, grasslands, wetlands and peatlands and

other carbon sinks, their biodiversity components and the loss of essential ecosystem services and leading to increases in greenhouse gas emissions due to these changes

2   Competition for land managed for the production of alternative crops, including land managed by indigenous and local communities and small-holder farmers, and competition for the commodity prices potentially leading to food insecurity

3   Increased water consumption, increased application of fertilisers and pesticides, increased water pollution and eutrophication, soil degradation and erosion

4   The uncontrolled cultivation, introduction and spread of GM organisms

5   The uncontrolled introduction and spread of invasive alien species

6   Emissions from burning biomass and potential adverse effects on human health.[41]

A major concern is the phenomenon known as the 'expansion of the agricultural frontier' which occurs when the increased demand for agricultural land means that forested and other uncultivated land is cleared for growing crops. Both sugarcane and palm oil production have been linked to the clearing of virgin forests. In Indonesia and Malaysia an estimated 14 to 15Mha of peat lands have been cleared for the development of oil palm plantations (Raswant *et al.*, 2008) although the oil is primarily for the food chain.

The direct land-use changes (LUC) that arise from the expansion of the agricultural frontier lead to environmental impacts such as loss of habitats, reduction in biodiversity, soil erosion, reduction in soil carbon, flood control and climate regulation. However, it is not only these environmental impacts arising from direct LUC that are raising concerns about increasing biofuel production capacities but the impacts that are likely due to indirect land-use changes (ILUC) (Koh, 2007; Righelato and Spracklen, 2007). ILUC occurs when the diversion of current food or feed crops (e.g. corn), or croplands (e.g. cornfields) to produce biofuels (e.g. corn-based bioethanol) causes farmers to respond by clearing non-agricultural lands to replace the displaced crops (Koh and Ghazoul, 2008).

Pressure from environmentalists has resulted in increased attention to the use of 'waste land' as well as promoting the drive to use second and third generation feedstocks. However, as was pointed out above, the term 'waste land' is contested, the absence of crops is not the same as providing no services or having any value to rural people (Clancy, 2008). Non-forest land not used for agriculture can support significant levels of biodiversity as well as playing a role in maintaining ecosystems by forming wildlife corridors and contributing to water resources (CFC, 2007). In addition, loss of access to this type of land and any associated reduction in biodiversity has social impacts on the rural poor, particularly women, where natural ecosystems form an important source of goods and services for rural households (Kartha and Larson, 2000).

Agricultural land is also an ecosystem but is generally not as rich in biodiversity as natural ecosystems. The type of agro-ecosystem also varies in its levels of biodiversity. Mono-cropping on a large-scale is generally considered negative for biodiversity as well as being prone to other environmental impacts such as soil erosion, soil fertility decline, pollution (Hartemink, 2005). These environmental issues raise concerns when thousands of hectares are converted to biofuels (see Chapter 4).

Doubtful climate change mitigation effects

There is substantial scepticism as to whether or not biofuels do reduce GHG emissions (see for example Searchinger, 2009; Lapola *et al.*, 2010). The values obtained depend on the feedstock, technology considered and boundary conditions assumed (UNEP, 2009). Inclusion of by-products as feedstocks can give a lower value for the GHG emissions per unit of output, whereas which GHGs are included in the analysis can increase the value; for example, ozone, a significant GHG, which is not a direct product of biofuel combustion but a product from secondary reactions, is not often included in assessments (Worldwatch Institute, 2007).

The LUC referred to above may in turn contribute to GHG emissions through the loss of carbon stored in above- and below-ground biomass when land is cleared; and/or the carbon sequestration service of converted land uses (e.g. growing forests) are lost (Worldwatch Institute, 2007). Peatlands and tropical forests represent large carbon sinks and if they are cleared to plant biofuels, the result is likely to be substantial emissions of soil carbon. It is estimated that globally peatlands contain about 528 gigatonnes of carbon, of which 42,000 megatonnes are contained in forested tropical peatlands of South East Asia (Hooijer *et al.*, 2006, quoted in Royal Society, 2008: 39). However, even if agreement cannot be reached on emission levels of GHGs due to LUC, or on approaches to measuring them, there are other environmental reasons for leaving these types of land untouched since they generally contain significant levels of biodiversity, which as we will show in Chapter 4 can be important for meeting the basic needs of the rural poor.

Nevertheless policy makers have responded to the criticisms related to biofuels and GHG emissions by incorporating the reduction of GHG emissions into sustainability standards. For example, the European Commission requires that the use of biofuels should result in a reduction of GHGs of at least 35 per cent compared to fossil fuels which increases from 2017 upward to 50 per cent for existing and 60 per cent for new plants (Jung *et al.*, 2010).

These climate change arguments are related to first generation biofuels – a reason for encouraging the focus on second and third generation biofuels (Commission of the European Communities, 2006; UNEP, 2009) which, as pointed out above, would not be to the advantage of the South.

Doubtful economic sustainability

It is difficult to give a generic statement about the cost of biofuels since it is dependent on a number of variables which have highly localised costs, including the type of feedstock, production efficiency and subsidies (FAO, 2008a). There is also a lack of data for countries other than within the EU and the US, as well as for 'less traditional' feedstocks such as jatropha (CFC, 2007). For example, bioethanol in 2006 was considered competitive with oil when the price of a barrel of oil was US\$70 for the EU, US\$50–60 in the US but in Brazil it was US\$25–30 (Dufey, 2006). The break-even price will not be static and will vary with both the oil price and the feedstock price. It will also vary over time in response to improvements in technology and infrastructure as well as institutional changes (FAO, 2008a).

A report prepared for the OECD has been most critical about the cost of biofuel production and has raised questions as to whether or not, with the current level of technology, they can be competitive with fossil fuels and other alternatives, such as oil from tar sands, without considerable government support (Doornbosch and Steenblik, 2007).[42] We do not wish to enter into a discussion of the rights and wrongs of subsidies, although we recognise the challenging issue of subsidies all too often being captured by politically powerful groups rather than the intended target group of low income households (CFC, 2007). However, relative to the oil industry, biofuels is new and there is some risk-taking involved in investing in biofuel production (not helped by oil prices fluctuating and uncertainties as to whether the automobile industry will opt for completely new technology, such as electric or hydrogen fuelled cars, or stay with liquid fuels). Research and development is needed for improving crop yields, improving conversion efficiencies and developing new technologies. It is not unusual under such circumstances for governments to provide subsidies for industries considered essential to the economy. It cannot be denied that Brazil would not be the market leader in bioethanol without subsidies to help establish and develop the industry. However, oil also is a beneficiary of subsidies. It has been estimated that in the US, in the 32 years prior to 2001, the oil industry received around US\$130 billion in tax incentives while the ethanol industry received US\$11 billion between 1980 and 2001 (Worldwatch Institute, 2007: 110).

If the objective of a biofuels programme is to reduce the costs of petroleum imports, there is a potential financial penalty for government revenues since any duty that would have been levied on the petroleum fuels is lost to the treasury. In São Paolo State, Brazil, the sum for forgone tax revenue as a result of bioethanol substitution for petrol was estimated at US\$0.6 billion in 2005 (CFC, 2007). The loss of substantial funds to the treasury reduces the funds available for state investment.

Impacts on agriculture

Others have questioned the wisdom of tying an agricultural commodity to oil prices (Kojima *et al.*, 2007). Energy markets are much larger than agricultural markets. A small change in energy demand can potentially create a large change in demand for agricultural products that can be converted into biofuels. This effect can work two ways. An increase in energy demand could divert agricultural feedstocks away from food markets and contribute to increased food prices, while a fall in energy demand could release an excess of feedstocks into agricultural markets, although it is uncertain whether this would mean cheaper food prices (at least in the short term).

When the demand for biofuel feedstocks is high and the price rises, this feeds through to other agricultural commodity prices since there is competition for the same resources. High prices do not necessarily reduce volatility in commodity markets (Doornbosch and Steenblik, 2007) (see Chapter 5).

High commodity market prices are considered to offer better income to farmers. However, income derived from higher prices is not necessarily transferred down the chain to the grower, particularly if the grower is a smallholder. Tenant farmers may not benefit from higher commodity prices since these tend to be matched by increases in rent for land.

There are also uncertainties about the reliability of the biofuels export market. Trade barriers in export destination countries could price biofuels in the South out of the market as happened to Pakistan (see above).

Human rights

Biofuels are becoming involved in some of the more unpleasant aspects of life in rural areas where in some places large-scale exploitation of natural resources can run hand-in-hand with displacement of rural people, land dispossession and violent appropriation or misappropriation of land. There are reports in Colombia of the involvement of paramilitary forces in displacing communities from their land, which is then appropriated for palm oil production (Christian Aid, 2009). The major expansion in production of palm oil in Malaysia and Indonesia has been criticised for being carried out with a disregard of indigenous people's rights, in particular when palm oil companies ignore, or at best do not understand, customary laws related to land rights. There is a well-documented land conflict in Sanggau district, West Kalimantan, Indonesia, that has been running since 1979 when palm oil was first introduced to the district (although initially biofuel was not the intended final product) (Cotula *et al.*, 2008).

There are also concerns about the labour conditions within large-scale agricultural plantations where there has been a history of poor working conditions, notably in the sugarcane and palm oil agribusinesses. There are reported incidents of intimidation of workers as well as the disappearance and murder of trade union officials in the palm oil areas of Colombia (Mingorance, 2006). The sugar plantations in Brazil were founded on slavery and unfortunately such conditions continue to exist on some estates. Child labour is also still a reality. It is difficult to give exact figures of how many children are involved because, unsurprisingly, official figures of what is an illegal activity are not available. The US Embassy in Brazil estimated (although no information is given how the statistic is compiled) in 1995 that approximately 3 million children between the ages of 10 and 14 were working in rural areas of Brazil, not all of whom were working on sugar/bioethanol estates.[43] In 2007 the Ministry of Labour in Brazil found almost 3,000 people, including indigenous workers, living and working under slavery-like conditions (Friends of the Earth, 2008).

While certification standards do incorporate human rights dimensions, the extent to which these are adhered to has been called into question (see for example Mingorance, 2006). There can be problems with independent verification of adherence to the criteria which can lack transparency (Buyx and Tait, 2011). Other financing schemes consider 'human rights abuses' are not part of their remit. An application for CDM (Clean Development Mechanism) financing by a Honduras palm oil company, which has been involved in land disputes and accused of human rights violations, was approved by the CDM Board. The decision was taken on the grounds that the Board was not 'equipped' to investigate the human rights abuses despite being presented with evidence about the abuses and the project partner who was to purchase the CDM credits withdrawing from the project.[44] There are also examples where women's rights are overridden. For example, where the male head of a household signs a deal with contractors who appropriate the land traditionally farmed by women for subsistence crops to feed their families and allocate it for biofuel crops (Vermeulen and Cotula, 2010).

This book

This book aims to contribute to the aspect of the debate on biofuels related to whether or not biofuels benefit the poor by examining the evidence from the literature. The definition of 'the poor' as a concept is ambiguous and the implicit meaning tends to reflect the scientific background of the user. For example, economists may use levels of income while social scientists may use a broad range of indicators such as health, education level. In this book, no rigorous definition is offered; instead when referring to 'the poor' the term is meant to encompasses rural dwellers not only with a lack of well-being (Dasgupta, 1993, quoted in Ellis, 2000: 77) but also whose assets – for example, of land, labour or cash – are at such a level that they are constrained to make investments in securing a sustainable livelihood,[45] including participation in biofuels projects (Reardon and Vosti, 1995; Ellis, 2000).

The pro-growth argument for biofuel production is based on the assumed stimulus that an indigenously produced product gives to the national economy not only through product diversification (in the case of biofuels) but also the foreign exchange benefits. The latter can accrue from direct substitution within the national fuel supply chain (hence reducing the oil import bill and saving forex) but also the potential for export earnings (hence generating forex). These are standard macro-economic arguments. However, what are the impacts at the micro-level where the effects on the poor are clearer? Dufey (2007) argues that biofuels may help improve rural employment and livelihoods. However, others consider that biofuels provide a considerable threat to the livelihoods of the rural poor. So do poor rural households stand to benefit from biofuel programmes?

We examine the evidence for the claims made by those who advocate for and against biofuels and attempt to answer the question: are biofuels inherently pro- or anti-poor? There is no universal agreed definition of what a 'pro-poor' intervention should consist of or result in (Hussain, 2004). There is considerable attention in development agencies on policies and programmes to create economic growth that is 'pro-poor' (see for example Ravallion, 2004). This approach has been one of the drivers for growing biofuel feedstocks as a stimulus to rural areas (see above). However, it is widely accepted that economic growth is essential but insufficient for poverty reduction (Chronic Poverty Research Centre (CPRC), 2008). Indeed, as will be explained in Chapter 2, poverty is no longer seen as purely a lack of sufficient income but it is multidimensional experience. To reflect a broader definition of poverty, in this book a pro-poor biofuels intervention is taken to be one that leads to better outcomes for the poor in terms of improvements in their assets and/ or capabilities.[46] There can be different forms of interventions: policy, institutional, managerial, legal or regulatory, financial, economic, infrastructural or technological (Hussain, 2004). The processes by which these interventions are identified and designed are also important for determining whether or not the outcome can be conceived of as pro-poor. It is also important to recognise that it is not necessary for the rural poor to participate in biofuel production chains for them to be affected, positively or negatively, directly or indirectly, by chain activities. Therefore, at the very least, the rural poor should not move deeper into poverty as a consequence of biofuel production.

Our approach is rooted in the Southern agenda outlined above, although this is inextricably linked to the Northern agenda. The focus is on the 'first generation biofuels' which are produced from sugar, starch and vegetable oils using conventional agricultural systems already embedded in the South. The processing technology is generally widely available and is used in developing countries. A switch to biofuel production may only require minor modifications to existing production plants. Biofuel programmes based on these technologies and existing production systems could, therefore, be implemented in the short to medium term, for example, in many Caribbean countries with economies dependent on sugar production under threat from reform of the EU sugar import agreements.

This book assesses biofuel programmes that are aimed at export markets, although there is some attention to local markets. We examine the hypothesis that growing and producing biofuels in rural areas can make a significant contribution to reducing rural poverty. We review the evidence on the changes that biofuel programmes initiate with particular reference to their pro-poor nature.

Chapter 2 describes the rural poor and gives an analysis of energy and poverty issues. The scene is set for the environment in which biofuels are expected to operate and provide opportunities for the rural poor. Chapter 3 then looks at the socio-economic impacts that large-scale biofuel productions are bringing to rural areas, in particular impacts on the assets of small-scale farmers and landless people in rural areas who rely on selling their labour.

The social impacts, particularly in terms of employment, the institutional issues related to biofuels and land tenure are then discussed followed by the gender aspects of biofuel programmes. Biofuels are analysed in relation to land-use changes at the local level in Chapter 4, in terms of their impacts on ecosystems and their services, and in Chapter 5 as a threat to food security together with the implications for the rural poor. Chapter 6, drawing on data presented in earlier chapters, looks at how biofuels as a fuel can address the energy and poverty issues identified in Chapter 2. The final chapter looks at the policy and institutional options for ensuring that biofuel production value chains are pro-poor, and finishes with some general conclusions about the institutional setting for a pro-poor biofuels programme.

2   Energy and rural poverty

This chapter introduces the rural poor and the context in which biofuels are expected to operate. It starts with some general statistics on poverty, and then it looks at what it means to be poor, particularly from the perspective of the poor. Understanding how people become poor and how they can move out of poverty is important for determining the potential impacts of new initiatives in rural areas such as biofuels. Such an analysis can help shape government policies towards biofuels where the goal is to address poverty. What is the best way to ensure that the rural poor benefit from biofuels development? Is it through participation in labour markets, hence the need to promote large-estate agribusiness production of biofuels? Or is it through smallholders participating in biofuel supply chains? Perhaps it is both of these, for different groups of rural people, possibly in different locations within the same country. Whichever form of production, in order to be pro-poor, poor people need to participate in biofuels supply chains on equitable terms of inclusion. At the very least, participation should not leave them any worse off than before. The chapter concludes by looking at biofuels not as a commodity but at the nature of the commodity, an energy source. Can biofuels benefit the rural poor in ways other than as a source of income? Energy and poverty linkages are analysed and the concept of energy poverty as a dimension of poverty is introduced.

The scale of rural poverty

Approximately 3 billion people live in rural areas which, at the end of the first decade of the 21st century, is just over half of the total world population. There is no standard definition of what constitutes a rural area (see for example the United Nations Statistical Division).[1] Two fundamental characteristics would seem reasonable: first, the major economic activity is agriculture and, second, the population density is low. The latter is relative; for example, the rural population density of the Netherlands is possibly greater than some small towns in the South.

The total rural population is expected to increase until around 2020 and then begin to decline both in absolute and percentage terms (except in Latin America and East Asia where it has already started to decline) (United Nations,

2008). As of 2002, the estimated number of the world's poor (when using the definition of people living on less than US$1/day) was approximately 1.2 billion of whom around 70 per cent are living in rural areas (CPRC, 2008). Over the period 1993 to 2002, the poverty rate[2] in developing countries declined from 28 to 22 per cent which has mainly been attributed to a decline in rural poverty. This decline has been linked to better living conditions in rural areas rather than to migration of poor people to urban areas. However, there are regional variations with the absolute number of poor people in South Asia and sub-Saharan Africa increasing since 1993 while China has made incredible progress with rural poverty falling from 76 per cent in 1980 to 8 per cent in 2001 (World Bank, 2007: 46).

Agriculture provides a means of living for an estimated 2.5 billion of the 3 billion rural inhabitants (World Bank, 2007). One and half billion rural inhabitants live in smallholder households (farming on 2ha or less) and 800 million work in smallholder households (World Bank, 2007: 29). Most of these households can be classified as poor, at least in terms of household income. These figures give some indication of the scale of the rural population dependent on agriculture that biofuels is intended to help. No one thinks that they can all be involved in biofuels production. The question is: how many?

In sub-Saharan Africa self-employment is the dominant labour pattern for the rural workforce, while in Asia and Latin America between 45 and 60 per cent of the rural workforce participate in labour markets (World Bank, 2007: 17). This division in employment patterns indicates that pro-poor biofuels policies might show significant differences between the three continents. With rising rural populations, declining farm sizes and slow expansion of agricultural employment, the absolute number of rural people living in poverty can be expected to increase, unless there are significant interventions to reverse the trend. As can be seen by the number of countries considering biofuel programmes (see Chapter 1), there is considerable anticipation that biofuels can make an impact and reverse the trend.

The confidence in biofuels is in part based on evidence that would appear to suggest that improving the lives of the rural poor has to rely on agricultural growth since there seems to have been a singular lack of success in transferring the economic benefits from growth in the non-agricultural economy to rural areas. It is thought that improvements in the non-agricultural economy have increased relative poverty, in terms of income levels, between urban and rural areas (World Bank, 2007). Economic growth from agriculture appears to be at least two and a half times as effective in reducing the poverty of the poorest quintile[3] as growth from other economic sectors (World Bank, 2007: 30).

It would, however, appear that diversification of income-generating sources in rural areas is occurring with a discernible shift from farm to non-farm income, even in places where there are good returns from agriculture for small-scale farmers (Rigg, 2006). This diversification includes many rural families, who might classify themselves as farmers, who have either out of choice or necessity diversified away from agriculture, to a range of activities such as trading and manufacturing in small-scale cottage industries. Young people in particular are making the transition away from farming. On the other hand, it appears that it is primarily the wealthier households that are benefiting from income diversification. The implications for pro-poor biofuels policy are that households with low incomes need support to be able to diversify into the non-agricultural parts of the production chain, such as processing and transport, so that these activities do not become the prerogative of the well-off. Young people might find the appeal of working with technology, or the upgrading of services (such as the arrival of the internet) in their villages sufficiently attractive that they may be prepared to stay in rural areas.

What does it mean to be poor?

Poverty is a multidimensional experience. People who live in poverty lack or have insufficient basic necessities, such as food, water, clothing and shelter to provide a sense of well-being, both material and psychological. They also lack other physical assets such as land, which can help provide some of those basic necessities. It is common to define poverty in monetary terms, for example, the well-known 'US$1 a day' measure of poverty used by international development agencies such as the World Bank. However, when people are asked to define what it means to live in poverty they tend not to refer to a single deprivation – lack of money – but instead talk about a range of resources they don't have without which their lives are harder and they feel vulnerable or lack respect (Narayan, 1999).

This is not to say that poor people set no store by money but that they see a lack of cash as only one among a number of deprivations. The non-financial resources are essential assets in their daily efforts to secure their livelihoods with the aim of moving out of or preventing the descent into poverty. These assets come in different forms: physical, human, social and natural (Ellis, 2000).

Land is an important physical asset which plays a multitude of roles. Not only does land provide a direct means of providing for a family but it can provide surety for credit and it can bestow status within the community. Poor people are not only deficient in terms of ownership of, or access to, physical assets but also in terms of access to those physical assets the state generally provides, such as roads, transport and modern energy carriers (particularly LPG and electricity).

People in a household are assets. The capacity to work is an essential ingredient in providing the necessities for well-being. This capacity to work requires, in the first instance, people to be healthy. This capacity can be enhanced through education and training, opening up the prospect for better paid employment and jobs with reduced drudgery. People are also part of social assets. They are the family and friends, as well as the network of other contacts established through participation

in organisations such as places of religious worship, farmers' cooperatives and women's clubs. These social networks help households to meet
everyday needs and provide the means – such as cash, equipment, new
skills and knowledge, and political contacts – and identify opportunities
either to prevent a household from entering poverty or to help it move
out of poverty. Networks play a particularly important role as a safety net
in times of vulnerability, such as disasters (natural and man-made), death
and macro-economic crisis.

Natural assets, such as soil fertility, water, plants and animals, depend on
the ecological system in which poor people live. Rural people, particularly
the poor, draw on these natural assets for a range of goods and services
(World Bank, 2008a). Poor people can be pushed, by large-scale commercial farming (including biofuel feedstocks), logging or mining, onto more
ecologically fragile systems such as steep mountainsides, where population
pressure can lead to a breakdown in traditional farming systems, in which
maintenance of the ecosystem has formed an important component. As a
result the ecosystem degrades and the natural assets decline in quantity and
quality with a corresponding decline in the availability of goods and services.
The impact of biofuels on natural assets has already been referred to in
Chapter 1 and will be discussed in more detail in Chapter 4. However, here
we can stress that biofuels can also deepen poverty or even move people into
poverty. This can happen as a result of large-scale agribusiness competing
with smallholders and other rural residents for natural assets, such as good
quality land and water.

Poor people are aware of their lack of influence over their own lives.
Their situation allows them to be treated without respect, particularly by
officialdom. They know that they are vulnerable to economic, social and
environmental upheaval which can deepen their poverty. Sometimes this
can manifest itself as a feeling of powerlessness to respond to expansion of
agribusinesses into their communities. On the other hand, poor people are
capable of organising themselves to oppose such expansion of biofuels (see
for example Friends of the Earth, 2008) or resistance in Latin America to
biofuels and international companies. The poor are subject to exploitation
by the more powerful in their own communities and to corrupt officials (see
Chapter 3). They also feel that they live on the edges of their society, often
not being able to take part in traditional ceremonies, rituals and festivals – the
very things that help create and cement their cultural and national identity
(Narayan, 1999).

Poor people see degrees of poverty based on levels of assets for those able
to work, and dependency for those who cannot. They are also aware of the
reasons for being poor. There is understanding, sympathy and, when possible,
assistance for those who are poor beyond their control: the elderly, widows,
the childless and the disabled. There is less sympathy for those who can work
but do not. In some cultures women who live without male support (unmar-
ried, divorcees and single mothers) can receive less sympathy since they are
considered responsible for their status (Narayan, 1999).

Poverty is considered a dynamic state, in the sense that households can move into or out of, or deeper into poverty. Most people who live in poverty work hard to try to improve their lives as well as their families'. A portion of people who live in poverty, however, are unable to work because of ill health, age (too old/too young) or disability. Indeed, the most overwhelming reason why households move into poverty is found to be illness (CPRC, 2008). A household member who is too ill to work reduces the income of the household and probably also increases household expenditure due to the need to buy medicines. Debt is another major cause of poverty. Poor people borrow money for a variety of reasons although health care, investment in improving production (such as irrigation), and meeting social obligations for weddings and funerals are commonly cited (Krishna, 2004).

Poor people are able to move out of poverty when they can accumulate enough assets to make the transition. Unfortunately a large number of people who live in poverty do so for a considerable part (if not all) of their lives. Why? There are a number of factors that prevent hardworking people from progressing. First, where they live is significant. The natural resource base on which many of them depend for their livelihoods is likely to be ecologically fragile (e.g. mountain slopes) or degraded through soil erosion and biomass removal (e.g. watershed[4] has been damaged). They may live in areas far from the main centres of political power. Second, even if they live close to political power centres, they probably have little political influence to help shape policies that could improve their circumstances. This lack of political influence is a factor in the lack of physical, communications and market infrastructure in more remote areas. Third, their social identity based on gender, ethnicity, religion and caste makes them vulnerable to exploitation (within households and communities). Fourth, the type of work they do depends on their human labour which is both low in productivity and costly in terms of time. These latter two factors are significant barriers to the accumulation of assets – assets that would enable the poor to have more secure lives and to survive many of life's daily problems, particularly to stay healthy, first through eating better and, when necessary, by affording health care. Time poverty is particularly a problem for poor women which means that most of their efforts go into survival activities to support their families, leaving little time for rest (important for well-being) and for participating in income-generating activities and community political, economic, social, and cultural activities – all of which can contribute to moving out of poverty.

Diversification of income-generating activities seems to be an important route out of poverty for both men and women. Moving to urban areas can be an option primarily for men. Others stay in their villages and take up new activities, both agricultural and non-agricultural. Biofuels are considered an opportunity for bringing diversity to rural areas. The crops can enable entry into a new value chain. Since first generation crops need processing close to the area where they are grown, new jobs can be created in the processing stages.

Who are the rural poor?

At the beginning of the chapter, the rural poor were closely identified with owning either small areas of land or none at all. However, the rural poor are not a homogeneous group, not only in terms of the extent of their poverty but also their reasons for being poor. These reasons can considerably influence a poor person's or household's capacity to participate in biofuels programmes and hence ability to move out of poverty. Where people live can be a significant determinant of poverty. Rural poverty has a distinct spatial characteristic: people living in remote areas, where the natural resource base is fragile and there is poor infrastructure, are prone to poverty (IFAD, 2010). Compared to non-poor households, poor rural households have more members, a greater share of non-working age members (i.e. dependants), less education, less land, less access to water and electricity (IFAD, 2010: 53). There are also specific groups (e.g. rural women, young people, indigenous people and ethnic minorities) who are disproportionately represented among the poor. Their lack of assets and poor capabilities, e.g. education, which leads to exclusion from opportunities that could move them out of poverty, arise due to power inequalities linked to identity in rural societies.

Women from poor households tend to be more disadvantaged than men from poor households in similar circumstances. For example, women's access to and control over assets such as land, cash and credit are more limited than men's. Women have less access to education, particularly beyond primary school, and skills development. Women's technical skills are often inferior to men's; for example compared to men, women's reading levels are lower and their experience with hardware is less. In particular women are 'time poor'. These disadvantages result in rural women having fewer options for earning income in skilled, rewarding and well-remunerated jobs. As a consequence, women tend to earn less than men. A study in 13 countries in sub-Saharan Africa, Asia and Latin America found that in nearly all cases, women's hourly wages ranged between 50 and 100 per cent of men's (Fontana and Paciello, 2009, cited in IFAD, 2010: 61).

Woman-headed households are particularly vulnerable to living in poverty. Women become heads of households for a variety of reasons, either through choice or through social circumstance (desertion, migration, divorce or widowhood). It is generally accepted that woman-headed households are more likely to be poor than man-headed households, although not all woman-headed households are poor (Narayan, 1999). As heads of households, women face an additional set of barriers to men in similar circumstances in coping with, and trying to move out of, poverty: when a family member falls sick they often have no other option than to act as nurse. They do not have the resources to hire someone to help either with household chores or to work on their farm plots (should they be fortunate to own or have access to this resource). The act of being the sole provider for their family prevents women heads of households from participating in new opportunities. There can be additional stigmas for single women, where in some cultures they are not able to compete on an equal footing even with other women, because they are regarded with suspicion to the extent that they will find it difficult to borrow money for investing in an enterprise (Narayan, 1999). There have already been concerns expressed about women's capacity to participate in biofuels production and the consequences for their participation (see for example Oxfam, 2007; Rossi, 2008). The concerns are based on wider experiences linked to agricultural production in general. These issues are discussed in more detail in Chapter 6.

Widows also face barriers to maintain their livelihoods due to cultural norms. Older women may be members of their children's household and provide childcare therefore they are unable to participate freely in labour markets. In many traditional rural communities a woman does not have the right to inherit her husband's property. As a consequence she is deprived of assets that could help sustain her livelihood. Physical assets such as land helps provide access to credit which might be required to help participate in new opportunities such as biofuel programmes.

Children are a group of rural dwellers who have little power or influence over their lives (Narayan, 1999). Children in poor households will more than likely suffer from disrupted schooling either because their families cannot pay the fees or they are required to contribute to the family's daily needs, such as by collecting water and fuelwood. Girls in particular are excluded from school to help with household chores. Poor households can disintegrate under social and economic pressures. Under these circumstances children might end up working, begging or be taken in by another family member, such as a grandparent. A growing phenomenon is the child-headed household as a result of war and HIV/AIDS leading to the death of both parents. The issue of child labour in large-estate cash crops, for example sugar, has been mentioned in Chapter 1. The causes of child labour are complex and there is considerable experience in how to offer alternatives so that poor families do not find themselves with the need for their children to work within and outside the home.[5] Biofuel programmes based on agribusiness need to ensure that these lessons are incorporated into their own employment practices.

Traditionally the elderly have relied on their children for support in old age. The elderly often play a vital role in childcare enabling parents to carry on with other productive tasks. If their children are poor, support for the elderly becomes increasingly difficult with advancing age and any kind of sickness. Social breakdown in some societies is leading to families neglecting social obligations to their elders who can rapidly become poor. Childless households can also face poverty even when they have land because there is no one to work the land. Biofuel programmes could look at creative ways to bring this land into production to benefit such households, for example, by providing labour to work the land although safeguards would be needed to ensure that elderly vulnerable people are not exploited. Other tasks, such as collecting fuelwood, can also be a problem with decreasing mobility, so the increased availability of alternative energy carriers might be particularly beneficial here (see next section on energy poverty).

Indigenous and tribal people form about 5 per cent of the world's population but are estimated to be 15 per cent of the world's poor (United Nations, 2009). Their poverty is linked to their livelihoods being based on natural resources over which they have precarious control. This is because their traditional land-use systems, governed by customary law, are not generally recognised under statutory law, which makes them vulnerable to the exploitation of natural resources by powerful outsiders, such as commercial logging companies.

There is also a social stigma attached to disability, which leads to the exclusion of disabled people from participating in labour markets. There is particular stigma for those diagnosed with HIV/AIDS. Ignorance around the disease continues to exist and people, once diagnosed, can lose their jobs, and their families also face discrimination. Without proper treatment, the capacity to work is lost far earlier than needs be. HIV/AIDS reduces the economically active population – a tragedy for the individual households as well as for the rural economy and society as a whole. Again, there are ways in which people with such challenges can benefit either directly, through appropriate forms of employment, or indirectly through improved services. The ethical dimensions incorporated into sustainability criteria for biofuels could provide a mechanism that the disabled and chronically sick benefit from biofuels programmes, such as improved health care.

Energy poverty

Energy is one of the most essential inputs for sustaining people's livelihoods. At the most basic level energy provides cooked food, boiled water and warmth. However, the energy carriers (the form in which energy is delivered to consumers) used to provide these services are strongly related to household income. Poor rural households primarily use biomass as an energy source and rely on their own metabolic energy to carry out all manual activities on the farm, as well as around the household and to transport goods to market. The source of biomass is the natural resource base of forests and shrubland where dry, dead fuelwood is scavenged, as well as on-farm resources of fuelwood, agricultural residues and animal dung. In many areas there is an increasing shortage in biomass supply, since the natural resource base has been degraded. As a consequence, biomass collection can take several hours per day, time that cannot be used for other livelihood activities. This situation was christened 'the other energy crisis' by the World Resources Institute as early as 1975 (Eckholm, 1975). Although nearly all households in rural areas use some biomass, poor households particularly rely on it, and tend to spend more time searching for biomass than higher income households. Wealthier households also purchase other, higher quality, modern energy carriers, such as electricity and LPG, which will be used for a greater variety of end uses than in poor households. In rural areas, poor households will generally restrict purchases for energy carriers to those of lighting (candles and kerosene – with their associated fire hazards) and batteries (small ones for torches and radios, and car batteries for entertainment systems). The fuel quality is low, burning with levels of smoke and particles that are recognised as having negative effects on health (see for example Smith, 1999)

Poor households use less energy per household than wealthier ones in absolute terms. Less water is boiled for drinking and other hygiene purposes, increasing the likelihood of water-borne diseases. This differentiation in quantity and quality of energy carriers related to household economic status has led to the formulation of a concept related to an energy dimension of poverty: energy poverty. Energy poverty has been defined as the absence of sufficient choice in accessing adequate, affordable, reliable, high quality, safe and environmentally benign, energy services to support economic and human development (Reddy, 2000). However, energy poverty receives little attention when addressing poverty issues in general. Energy is by and large not recognised as a major factor in addressing poverty as can be seen by the failure to mention it specifically in the Millennium Development Goals (MDGs).[6] The IEA considers that implicit in meeting the MDGs is the need to provide over 550 million people with access to electricity by 2015 (OECD/ IEA, 2010).

Why do people live in energy poverty? There are two contributing factors: availability of good quality energy carriers together with their associated energy efficient technologies, and the capacity to pay for them. Modern energy carriers are based on capital intensive production and distribution networks of grids and pipelines and transport systems. The cost of delivery increases with distance. Unfortunately for the rural poor they often live in the more remote areas with the sort of terrain, such as mountains, that increases the costs of putting in the distribution networks. The population density in rural areas, where people tend to live in small, scattered settlements, also increases distribution cost. Supply companies do not see rural areas as a good return on investment and are reluctant to supply such areas. As a consequence, there is limited availability of modern energy carriers in rural areas.

The capacity of rural households to pay for energy carriers and the associated energy conversion equipment is limited. This situation results in continued 'low productivity, low quality of outputs, and an inability to release labour for economic activity. In turn, this leads to low returns on investment and labour inputs, again limiting the capacity to acquire modern energy services and appliances' (Ramani and Heijndermans, 2003: 19). As a consequence, poor households become trapped in a vicious circle of energy poverty. However, savings on energy expenditures can help move households out of poverty. A survey in Sri Lanka in the late 1990s found that savings of Rs100 (approximately US$0.75) on kerosene would have moved 22 per cent of the poor households surveyed above the poverty line. Such savings were estimated to be 12 per cent of the monthly income (University of Reading, 1999).

Ending energy poverty requires enabling access to modern, cleaner energy carriers, e.g. electricity, LPG and biogas, or more efficient conversion devices for biomass, such as improved cook stoves. An estimated two billion people are without access to clean energy carriers. These modern forms of energy are regarded as an essential input for economic development, since they substitute for human and animal energy and hence increase output, reduce time poverty, provide healthier working conditions, as well as providing new services, such as refrigeration, not only for the household but also the community (Barnett, 1999). Energy alone, however, is not sufficient to bring development because other complementary inputs, such as roads and finance, are also required. Indeed, it may be more efficient to 'bundle' energy provision with the provision of other services: water, sanitation and education. For example in Peru it was found that the addition of a fourth service for rural households has had an effect seven times greater than the addition of the second service (Barnett, 2000).

At the household level, people do not want energy carriers as such but the services, such as heat and light, which these energy carriers can provide. Again at the household level, complementary inputs in the form of conversion technologies are necessary to obtain the desired energy services. Access to energy services provided by modern energy carriers can bring about an improvement in the living conditions of the user. For example, access to electricity provides better quality lighting as well as providing new services such as telecommunications. For poor people, these new or improved services contribute to poverty alleviation although it may be difficult to actually attribute such effects solely to energy since there are a host of other factors, such as macroeconomic policy, which affect measurable poverty (IDS, 2003). However, if people are to move out of poverty, they need to be able to use modern energy carriers to raise incomes through increased farm output and improvements in quality, as well as diversification in the types of income-generating activities, both on- and off-farm. Therefore, by addressing energy poverty, overall poverty can be reduced. The impact on poverty of improved energy services is determined by the choice of end-use to which energy is put.

The gender dimension of energy poverty

While both men and women benefit from access to energy in terms of reducing poverty and hunger through increased food production, employment and clean water, women and girls are likely to show additional benefits due to time saving, particularly in terms of water and fuelwood collection, and improved health, particularly through the use of cleaner energy carriers (Ramani and Heijndermans, 2003).

The gender dimension of energy is based on the gendered division of labour within households which generally allocates to women the responsibility for household energy provision (Moser, 1993). They are often supported in this work by girls and sometimes boys, who can be kept out of school to help with the search for biomass thereby damaging their own future livelihood choices (Agarwal, 1997; Vogt, 2007). Providing energy to meet household needs can mean spending many hours collecting fuelwood. For example, Malmberg Calvo quotes figures of more than 800 hours a year in Zambia and about

300 hours in Gambia and Tanzania (Malmberg Calvo, 1994, quoted in Rossi and Lambrou, 2008: 10) The loads carried can weigh 20kg or more and do cumulative damage to women's spines and internal organs (see for example Amacher *et al.*, 1993; Cecelski, 1995; Wickramasinghe, 2001; Cooke *et al.*, 2008). Men do get involved in fuelwood collection although the trigger for their involvement appears to be dependent on local circumstances, such as fuelwood scarcity (Cooke *et al.*, 2008). Women are also responsible for household tasks that ensure the survival of the family, such as preparing food and providing water. Cooking with fuelwood results in women and young children, who are often carried on their mother's backs or are kept close to the fire for warmth and supervision, being exposed to high levels of smoke for between three and seven hours per day (WHO, 2005). This type of exposure has been attributed to the higher levels of lung and eye diseases suffered by women compared to men (Smith, 1999).

Cooking with solid fuel is considered responsible for 1.6 million deaths due to pneumonia, chronic respiratory disease and lung cancer. The World Health Organization considers that 59 per cent of all female deaths are attributable to indoor air pollution and it also accounts for 56 per cent of all deaths in children under five years of age. As well as fuelwood collection, fetching water and grain preparation are particularly demanding on women both physically and in terms of time spent on such activities (Cecelski, 1995; SANDEE, 2007).

Other gender issues which influence the household's type of energy carrier and how it is used relate to women's influence over decision making within the household and community. Women have in general less influence than men over decisions and exercise less control over their own lives and resources, both at the household and community levels (Moser, 1993). Consequently, women's capacity to control processes and resource allocation on many issues, including energy, is limited. Men and women often use, are impacted by, or benefit from energy services differently. The same energy service may have different social or economic outcomes for men and for women. For example, men may choose to locate a light outside the house for security reasons (such as protecting livestock from theft) while women may choose to locate the light in the kitchen (Cecelski, 2000). Women and men have different perceptions about the benefits of energy. For example, a research study on the gender-related impact of micro-hydro in Sri Lanka, found that men in the area under study saw the benefits of electricity in terms of leisure, quality of life and education for their children; women saw electricity as providing the means for reducing their workload, improving health and reducing expenditure (Dhanapala, 1995, quoted in Barnet, 2000).

Rural energy needs

Rural energy needs can be classified into household, income-generating (agriculture and small enterprises) and community services. Table 2.1 shows, for three broadly defined income groups, some typical end uses and the energy carriers used to meet these needs. Two important points to note are that, first, low income[7] households use mainly biomass and their own labour; second, expenditure on an energy service is usually for lighting.

At the household level, energy can play a role in combating rural poverty through:

1   improved health
2   increased productivity and new opportunities for additional income
3   reduced labour and time spent on household activities.

Table 2.1 Typical end uses and their energy carriers differentiated per income group in developing countries

Household	Income level		
	Low	*Medium*	*High*
Cooking	Wood, residues and dung	Wood, charcoal, residues, dung, kerosene, biogas	Wood, charcoal, kerosene, LPG, coal
Lighting[8]	Candles, kerosene, none	Candles, kerosene	Kerosene, electricity
Space heating	Wood, residues, dung, none	Wood, residues, dung	Wood, residues, dung, coal
Other appliances: radio/television	None	Grid electricity and batteries	Grid electricity and batteries
Space cooling and refrigeration	None	Electricity (fans)	Electricity, kerosene, LPG
Agriculture			
Tilling	Human labour	Draft animals	Animal, gasoline, diesel,
Irrigation	Human labour	Draft animals	Diesel, grid electricity
Processing	Human labour	Draft animals	Diesel, grid electricity
Industry			
Milling/mechanical	Human labour	Human labour, draft animals	Grid electricity, diesel, gasoline
Process heat	Wood, residues	Coal, charcoal, wood and residues	Coal, charcoal, wood, kerosene, residues
Cooling/refrigeration	None	None	Grid electricity, LPG, kerosene
Services			
Transport	Human labour	Draft animals	Diesel, gasoline
Telephone	None	Batteries	Grid electricity

Source: World Bank, 1996: 25

The categories are linked; for example, the reduced time spent on back-breaking physical labour allows the body recuperative time. In the first category, the use of biomass for cooking contributes significantly to indoor air pollution, and shortage of biomass can lead to a reduction in boiled water which has implications for prevention of the spread of water-borne diseases as well as general hygiene. Here is the need, therefore, for both an increase in the availability of energy carriers and cleaner combustion. Improved health can be achieved by prevention and by cure. Having sufficient good quality food is always a challenge for the poor. Improved crop output for those with access to land or increased incomes to buy food can help. Agricultural production can be improved through energy services, based on mechanisation using diesel engines and electricity. Increased output, improved processing and storage means not only more food for the family but potentially a surplus to sell, which increases household income. Mechanisation also reduces drudgery which has positive health benefits. Increased income means the household has the financial assets to pay for any necessary health care.

As was pointed out above, there is increasing diversification of the basis of rural livelihoods. There are large-scale agricultural estates and their associated processing industries, such as sugar and palm oil, and mining providing employment opportunities. However, many livelihoods are based on small- and medium-scale enterprises such as grain milling, fruit and vegetable processing, charcoal making, brick burning, potteries, bakeries, blacksmiths and carpenters. The energy service needs of rural industry are for lighting, process heat and motive power. Lighting is provided by kerosene or grid electricity, while biomass produces process heat and motive power comes from human and animal labour, diesel engines and grid electricity.

Where income-generating activities are centred on the physical structures of the household, such as food preparation and sewing, lighting can play an important part in extending working hours.[9] In this respect, households of all income categories consider electric lighting to be the most desirable benefit (Ramani and Heijndermans, 2003). Electricity can make a contribution to income generation by allowing access to information through radio, TV and reading and increased comfort through the use of fans. The major benefit of the provision of grid electricity appears to be a reduction in women's time poverty. A study in Sri Lanka found that 80 per cent of the interviewees save between one and two hours a day through avoided journeys (taking batteries to be recharged, and going to the city to buy kerosene, medication and vaccinations) and on household activities (such as firewood collection, cooking, ironing, boiling water, house cleaning and chimney cleaning) as a result of having access to grid electricity (Massé and Samaranayake, 2003).

However, to address other aspects of time poverty requires services, such as water-pumping and grain milling, to be provided at the community level. A survey in northern Tanzania found that electric grain mills saved women and children considerable amounts of time in walking to and waiting for grain to be milled. The saved time has been used for other income-generating activities or studying (Clancy and Kooijman, 2006). Health clinics, schools and sanitation are part of community services and make a significant contribution to addressing core issues of poverty: poor health and illiteracy.

In terms of income generation, the rural poor are limited in their options because other input factors, such as investment capital and access to markets, are not available. People with limited assets are often risk averse and so they are reluctant to take decisions that might tie up labour, land or financial assets (Will, 2008). Those with land are able to engage in some agricultural production (although they may not always produce enough to feed their families), animal husbandry (which often relies on access to common land) and, in some places, fishing. The landless must rely on selling their labour. They can hope to benefit indirectly from energy services when larger landowners make use of irrigation, postharvest processing and storage. On the other hand, access to modern energy services can result in mechanisation which leads to job losses.

Energy and rural poverty

This chapter has set the scene in rural areas where biofuels will have to operate. It has also indicated issues related to the rural poor which need to be taken into account if the rural poor are not to be excluded from the benefits of biofuels development. There is clearly a need in rural areas for initiatives that will address a number of issues related to the rural economy such as the growing population. Agricultural initiatives have been shown to improve the living standards of the poor. An argument for biofuel projects is that smallholders can benefit by growing biofuel crops and the landless can participate in labour markets. Stimuli to rural economies can also come from non-agricultural activities such as biofuels processing and providing goods and services for rural communities. The latter can develop where there are significant numbers of wage labourers who can migrate into an area where new large-estate grown biofuels are established. It was pointed out in this chapter that it tends to be the already better-off households that are able to diversify their income sources and participate in new initiatives. Therefore, to be pro-poor, biofuel initiatives would need to be formulated to ensure that less well-off households (both with and without land) and those where labour availability is constrained are able to participate. To what extent this is happening is reviewed in Chapter 3.

The poor are not a homogeneous group in terms of either their social characteristics or their reasons for being poor. It is quite possible that some of the poor will not directly benefit from biofuels coming to rural areas because they are not able to work: for example, they are too young, too old or too ill. This group can benefit if there is general improvement in the level of rural services such as well-equipped health centres and schools. There are other groups of poor people who do not have sufficient assets or face social constraints to participate in growing and processing biofuel crops. Women in particular can be disadvantaged because they often face more constraints than men from the same social group. Groups that are asset poor may need specific targeted initiatives to ensure that they move out of poverty as a consequence of biofuel programmes.

Poverty is now seen as more than a lack of income, although having access to cash is important for providing necessities such as food, clothing and medicines as well as to pay school fees. People are often poor because of where they live and who they know. Remoteness in terms of geography can mean living in areas that would not be attractive for large-scale growing biofuels; for example, transporting the products may not be economic and roads may be inadequate, although circumstances may allow for small-scale production for local consumption. The lack of physical infrastructure in rural areas can be linked to remoteness from the centres of political power. Poor people have little political influence which could be significant in determining whether or not biofuels come to an area or that the policy framework ensures the poor get an equitable deal. Indeed, as will be seen in Chapter 3, the rural poor can be marginalised by the authorities when large-scale agribusiness moves into an area.

Poor people lack other sorts of assets, such as land, that can prevent their participation in biofuels projects. Land plays a dual role, not only for growing crops but also as leverage for investing in new initiatives. Women can be significantly disadvantaged by their lack of control over land. Knowledge and information about biofuel initiatives is an important component in participation. Social networks are considered particularly significant in providing information, especially when the networks extend into urban areas. However, the poor tend to have weak social networks.

This chapter has also described two significant aspects of poor people's lives: energy and time poverty. These two concepts are interlinked. The poor, particularly women, are time poor because they generally rely on their own labour for many of their everyday tasks that occupy most of the daylight hours. They are energy poor because they generally rely on their own labour (and sometimes animals) and biomass (the collection of which contributes to time poverty) and other low quality energy carriers. Energy poverty exists because there is usually limited availability of better quality alternatives. Where alternatives, such as electricity and LPG, are available they often have high up-front costs and the poor do not have the capacity to pay for them. A vicious circle exists in which a lack of time and lack of energy carriers, which can increase productivity, result in a lack of accumulation of capital, which would allow for the purchase of higher quality energy carriers. Biofuels can play a multiple role in breaking this vicious circle. First, the value production chain opens up the opportunity for improved incomes. Second, if such projects bring improved infrastructure and larger markets the availability of modern energy carriers may improve. Third, biofuels are energy carriers. Their production in rural areas allows for the possibility of decentralised energy systems serving local communities, as already happens on estates growing crops such as sugar and tea.

Urbanisation, linked with increased income and purchasing power in parts of the South, has had an influence on lifestyles in urban areas, the impacts of which are felt in rural areas. There has been a significant increase in the number of motor vehicles in use, particularly in India and China,[10] which increases the demand for liquid fuels. Indeed, the sudden surge in oil prices in 2007–08 to more than US$100/barrel has in part been attributed to the increased demand from China.[11] The oil price rise has been one of the driving forces for the search for alternative transport fuels, including biofuels, in the industrialised countries (Mitchell, 2008). Another significant transformation in urban lifestyles has been a change in diet patterns with a move away from starchy foods towards more meat and dairy products, which in turn has increased the demand for grain to feed livestock (FAO, 2008b). So it is not only biofuels but also shifting patterns of agriculture that are bringing transformations in rural areas. The environmental impacts, in particular on land use, of these transformations are discussed in Chapter 4, while the effects on food security are discussed in Chapter 5.

The chapter started with some statistics about the dynamics of rural populations. A long-term decline in the rural population in the South is predicted both in terms of the total numbers of people living in rural areas and the ratio compared to urban populations. These apparently rather sterile figures have practical realities for rural areas and rural livelihoods. Urbanisation might bring better opportunities for rural migrants in terms of access to infrastructure and educational opportunities, but for those left behind the situation does not improve, although remittances from migrating family members may bring some respite. Absolute and relative poverty in rural areas continues to be higher than in urban areas (World Bank, 2007). The need to reduce both absolute and relative poverty is one of the factors driving policies for growth in the rural economy. In this context, biofuels are clearly seen by some as providing an opportunity for stimulating rural growth and farm incomes (UNDP, 1995; Kammen *et al.*, 2001; Kartha *et al.*, 2005; World Bank, 2007). However, to what extent do biofuels production chains really provide an opportunity for the rural poor? Are they able to become involved as owners of part of the production chain and able to influence the governance of the chain, or are they marginalised as paid labour working for large-scale agribusiness? In this context, the next chapter looks at the prospects for the poor to participate in biofuel production chains; while Chapter 6 looks at whether or not biofuels are able to contribute to addressing the energy and poverty issues raised in this chapter.

3 The impacts of large-scale liquid biofuel production in rural communities

Large-scale biofuel production offers the prospect of product diversification based on either traditional or new crops. Selling a crop to a new market can help spread the risk of price fluctuation on commodity markets and it can provide an outlet for surpluses. These opportunities are seen as particularly important for many of the traditional sugar-producing countries of the Caribbean and Africa since they offer a means to compensate for the lost revenue due to the loss of preferential quotas and a 36 per cent reduction in guaranteed prices under EU sugar reform.[1] There is also the prospect of refining the biofuels in rural areas, hence the value added by converting the raw material into the final product remains local.

The delivery path for large-scale production can be on the basis of agribusiness plantation grown crops using wage labour or a central processing plant based on outgrowers or a mixture of the two. Refining first generation biofuel crops needs to take place close to the growing sites since the biomass material generally begins to deteriorate rapidly after harvest.

This chapter looks at the socio-economic impacts that large-scale biofuel production is bringing to rural areas, in particular impacts on the assets of small-scale farmers and landless people in rural areas who rely on selling their labour, as well as the distribution of benefits in respect of gender. The focus is on biofuels for export markets, while Chapter 6 looks at the possibilities for serving local markets. One of the major criticisms directed at biofuels has been the vulnerability of the poor to rapid expansion by large-scale biofuel programmes, in particular how this expansion affects their access to land. This chapter, therefore, examines the institutional issues related to biofuels and land tenure.

The ways in which poor people in rural areas can benefit from biofuels is either by direct production for the value chain or through paid employment. The first section in this chapter will concentrate on the former, and the following section on the latter.

Income generation

Not everyone opposes development of biofuel production in their community. Villagers can see the arrival of outsiders to create large-scale schemes as bringing new opportunities, such as income generation through job creation and a local market for their own products and services. Combining growing biofuel crops with existing export crops can offer diversification, which reduces vulnerability to price fluctuations and sudden policy changes in the export countries. Volatility in commodity markets has been one of the reasons why small-scale cotton farmers in Mali have started to grow jatropha in place of cotton (Practical Action Consulting, 2009). Biofuel developers may also offer other incentives, such as technical assistance and seeds.

In terms of the organisation of value chains, large-scale producers are considered to have advantages both in terms of economies of scale and logistics, as do vertically integrated agro-industrial chains (UN-Energy, 2007). However, in biofuel production, economies of scale are considered less important in feedstock production than processing (Peskett *et al.*, 2007a). Small-scale farmers are capable of producing at levels of efficiency close or equal to large-scale producers. For example, small sugar farms in Thailand compare favourably with large and medium-sized farms in Australia, France, and the US (Larson and Borrell, 2001, quoted in Kojima and Johnson, 2005: 100). There are fears that small-scale producers will be under pressure to take lower prices for their feedstock as refiners are under market pressure to reduce their costs (Peskett *et al.*, 2007b).

Are particular biofuel crops more likely to favour smaller farmer production? A number of biodiesel crops, most notably jatropha and palm oil, are considered best harvested by hand and hence are suited to small-scale cultivation. Smallholder participation in palm oil is already extensive, for example, in Indonesia smallholders account for around one-third of national annual production (Vermeulen and Goad, 2006). It has been suggested that jatropha may be more economically and environmentally viable in small-scale community-based jatropha plantations, agroforestry systems with jatropha intercropping and agro-silvo-pastoral systems than large-scale monoculture plantations (Achten *et al.*, 2008). Sweet sorghum is being considered in Africa since this is similar to the grain sorghum grown in many countries as a staple crop on smallholdings. Its cultivation practices, and water demand, are considered appropriate for the region (ICRISAT, 2007). However, bioethanol crops such as sugarcane and maize are usually grown on large plantations to allow for mechanisation of harvesting and hence reduce costs through economies of scale. Outgrower schemes do exist for small-scale farmers to grow sugarcane, for example, in Kenya and South Africa (Cotula *et al.*, 2008).

There is no guarantee that a crop grown as a monoculture on a large scale will do equally well on a small scale. *Campesinos* in Paraguay have struggled to get good yields from soy on their smallholdings[2] (Palau *et al.*, 2008). This is increasing their indebtedness to the seed merchants from whom they buy the genetically modified seeds and other inputs such as herbicides. To increase their incomes some farmers are displacing their subsistence crops which increases their vulnerability to food markets. The *campesinos* have switched to soy because the price of cotton is no longer profitable. Soy requires less labour intensive cultivation than cotton. While it might at first sight seem a blessing that people no longer have to work long hours in the field, this is only the case if there is alternative employment to absorb the now surplus labour. The consequence is an outward migration of labour; while this might improve household financial assets through remittances it puts the family as a social unit under strain. The demographics of rural areas can also change if it is primarily the young who leave.

Since many of the first generation biofuel crops have to be processed within 24 hours they cannot be transported across large distances for central processing which does limit the scale of process facilities (Vermeulen and Goad, 2006). An appropriate crop, therefore, opens up opportunities for small-scale farmers to sell to local processors. If refining into biodiesel is also carried out at this scale, a significant portion of the value added has been captured at the local level. If smallholders are to participate in biofuel value chains, coordination of supply to a central processing facility becomes a key issue, since large-scale processing of biofuel requires a steady flow of feedstock to ensure a steady output to meet contracts. However, in the Brazilian ethanol programme in the 1980s, schemes set up to benefit small-scale farmers as growers of cassava supplying a central processing facility were not generally successful in terms of the logistics of supplying the raw material to the plant, which proved to be too complex (Clancy, 1991). Teething problems have been reported with the Brazilian Biodiesel Production and Use programme (PNPB), in which farmers who were members of a cooperative chose to sell their castor beans to international commodity markets where the price was more favourable than that offered by Petrobras with which the cooperative had a contract. The cooperative was unable to meet its contractual obligations and was fortunate not to face a fine (Bailey, 2008).

There are a number of forms that delivery paths can take, including outgrower schemes, cooperatives, marketing associations, service contracts and joint ventures. There are examples of small-scale growers' coopera-tives offering an alternative model to large-scale plantations (see Box 3.1). There is a body of experience with different forms of participation by small-holders being built in the palm oil industry in South East Asia (although not specifically for biofuels) (Vermeulen and Goad, 2006). The effect is more pronounced for producers who receive support to improve their production than independent producers. Producers with links to refiners have the advan-tage of a guaranteed market but independent growers have flexibility in deci-sion making. In particular, independent growers have the potential to gain a greater share of the value chain of palm oil because they can be owners of processing facilities. Ownership, particularly of the downstream components of the value chain, is considered essential in terms of realising the maximum benefits from the value chain (CFC, 2007). It allows for a diversification of income and a buffer against low agricultural prices. In terms of rural devel-opment, it has been estimated in the US that the value of a 40 million gallon bioethanol plant is about US$1.5 million a year if the facility is owned by

---

Box 3.1 Small-scale biofuel crop production: an alternative approach to agribusiness

The Coopaf cooperative in the north-east of Brazil has 5,000 members many of whom are descendants of escaped slaves. Small-scale farmers face challenges to secure a reasonable income due to the semi-arid conditions and a lack of infrastructure. However, the Brazilian government's biodiesel programme has begun to make a difference to their income. Cooperative members have responded by growing castor, intercropped with corn and beans. Many of them also grow vegetables and keep livestock. Castor appears to be a good crop for diversification since it requires only one month of rain, which fits well with the decline in rainfall that has been experienced over the last 12 years. Beans require three months of rain, therefore castor can create a greater degree of security. The advantage of being a member of the cooperative is the technical assistance that farmers receive as well as having access to a fixed price guaranteed by the biodiesel programme. In 2007, the programme paid 36 reais (~€15) for a 60kg bag of beans compared to 12 reais (~€5) that farmers received before the programme. For the next season the cooperative has been able to negotiate a price of 45 reais (~€18.5).

(Source: Bailey, 2008)

---

external investors, and US$6–12 million a year if the farmers own the plant (Morris, 2006: 3). On the other hand, multiple ownership of a facility has proved difficult in terms of raising capital and producers can be reluctant to put up their land certificates as guarantee (Vermeulen and Goad, 2006).

If smallholders are not to own the processing facilities, then the agreements between them and the biofuels refiners need to ensure an equitable distribution of the revenue. There are different forms of such schemes with different degrees of success. On the positive side, in Madagascar, the use of 'micro-contracts' and the provision of support and supervision for small-scale farmers (albeit producing for the export food market rather than biofuels) has been found to result in higher welfare and greater income stability (Worldwatch Institute, 2007). Contract farming schemes offer price stability for small-scale farmers, although if market conditions change over time, the contracts can become less attractive and run the risk of being seen as unfair. In Namibia, the Kavango Biofuel Project uses a joint venture between a farmers' association and a local company. Farmers contribute communal land and labour, while the company covers capital costs and compensates participating farmers with food and cash for loss of maize and millet. Not everyone in the community is able to contribute land to the project, so to try to ensure equity of access to benefits, the project plans to compensate these community members by giving them priority for other project-related employment opportunities (e.g. tractor drivers, factory employees) (Jull *et al.*, 2007). Box 3.2 gives examples of other schemes intended to benefit small-scale farmers.

Box 3.2 Examples of schemes intended to benefit small-scale biofuel producers

In India, the International Crop Research Institute for the Semi-Arid Tropics (ICRISAT) supports a public–private consortium aimed at testing and growing sweet sorghum. To meet its target for ethanol production, the private Indian company Rusni Distilleries contracted approximately 3,200 small-scale farmers to grow sweet sorghum on roughly 2ha of land each. Kaveri Seed Company provides high quality seeds to the farmers, while ICRISAT contributes research inputs and technical advice.

(Source: ICRISAT)

In Mali, Mali Biocarburant SA – a private company which has received financial support from the Government of the Netherlands to cover business development costs – is producing biodiesel for the national market from *Jatropha curcas* without acquiring land and developing plantations. Small-scale farmers from Mali are shareholders in the company. They supply the jatropha nuts to the Union Locale des Sociétés Coopératives des Producteurs de Pourghère à Koulikoro (ULSPP), a farmer association which extracts the oil and sells it to Mali Biocarburant. The seed cake produced as a by-product of the oil pressing is sold to the farmers to improve soil fertility. Mali Biocarburant then processes the oil into biodiesel and sells the by-product (glycerol) to a women's cooperative to produce soap. A private company, Interagro, purchases the fuel. Farmers not only earn revenue through the sale of the nuts, but also through dividends and increased share value.

(Source: Mali Biocarburant)

In Tanzania, the Tanzanian FELISA company – funded by equity contributions mostly from Belgian shareholders – targets 10,000ha of land for oil palm plantation. To date it has acquired 4,358ha, set up a large oil palm nursery (42,000 seedlings), installed processing equipment and mobilised 990 farmers to participate in the scheme as outgrowers. The farmers received 10,000 free seedlings and were trained in palm husbandry. The farmers are under no obligation to sell only to FELISA, and the price is negotiable; however, the contractual agreement may bind them to supply a certain amount of a crop at a specified quality over a given period of time.

(Source: FAO and PISCES)

In Uganda the International Fund for Agricultural Development (IFAD) aims to increase small-scale farmers' income by revitalising national vegetable oil production from oil palm. Implemented in partnership with a private sector company, Bidco Oil Refineries, it targets an area of 10,000ha of land located in Bugala Island, Kalangala District. About 3,500ha are cultivated by 1,400 smallholder farmers through outgrower schemes. IFAD's funds supported the establishment of Oil Palm Uganda Limited (OPUL) – a consortium in which Bidco and the small-scale producers are partners – and the Kalangala Oil Palm Growers Trust – the local farmers' association which has a 10 per cent share in OPUL. The trust provides farmers with credit and helps them to obtain fair deals when selling their produce. OPUL provides seedlings and fertilisers, technical support, housing and meals to its employees. It also built roads and runs a clinic.

NB There are no independent evaluations of these schemes.

(Adapted from Haralambous *et al.*, 2009)

Whatever the form of agreement, given the imbalance in experience and expertise available for negotiating between small-scale farmers or communities and international biofuel investors, institutional support needs to be provided to farmers to assist with contract negotiations (see below). The PNPB in Brazil provides this type of support. On the negative side, refining companies have not honoured contracts with small-scale farmers (see for example in Indonesia (Oxfam, 2007)). Indeed, in the palm oil sector, the use of detailed written contracts for outgrowers is considered less common than in other value chains, which may make smallholders vulnerable to exploitation (Vermeulen and Goad, 2006).

Infrastructure also plays a role in who benefits. In the palm oil agribusiness in Honduras it is estimated that 80 per cent of producers have no access to transportation and so they are obliged to sell to intermediaries who collect the palm fruits. As a consequence, they have little influence with the extractor companies and even risk exclusion from participation in an international supply chain (Fromm, 2007, quoted in CFC, 2007: 57).

Employment

As was shown in Chapter 2, the rural poor are not a homogeneous group and their livelihoods are likely to be affected by biofuels programmes in different ways. The previous section looked at how small-scale farmers can benefit from biofuels programmes. However, not all are likely to be so fortunate. Small-scale farmers, particularly those who do not have official title to their land, are potentially vulnerable to being displaced by consolidation of landholdings. In part this is driven by mechanisation to keep costs of biofuel feedstock production down. In São Paulo State mechanisation will be obligatory by 2017 (Bailey, 2008) so the number of jobs can be expected to decline further. This means that displaced farmers and their families will be competing with those rural people already looking for non-farm employment. Can these people be absorbed into the agribusiness created by biofuels? Biofuel production is predicted to generate more employment per unit of energy than conventional fuels and more employment per unit investment than in the industrial, petrochemical or hydropower sector (UN-Energy, 2007). Dramatic claims (such as 'more than one million jobs') are often made for the numbers of jobs created in the Brazilian fuel ethanol programme (Kartha and Larson, 2000). It is contested how many of these jobs are actually new ones. Indeed the accuracy of counting has also been called into question. An analysis of the claims for the number of jobs created in the state of Iowa, USA, finds that there is a tendency to overestimate the number of jobs, particularly indirect ones, by as much as tenfold (Swenson, 2006). In São Paulo State, where the most modern production facilities exist converting one million tonnes of sugarcane to ethanol, the estimated employment generation is 2,200 direct jobs (1,600 in agriculture and 600 in production) with 660 indirect jobs in service support (Kartha *et al.*, 2005). Considering the area of the state is approximately 70,000ha, the impact of the spending power of these employees is highly localised. There has also been an increase in mechanisation which has brought improvements in the form of increased wages and improved labour conditions but at the same time reduced the numbers of jobs. A measure of the impact of mechanisation is that total employment in Brazil's sugarcane industry has declined from 670,000 in 1992 to 450,000 in 2003 (Worldwatch Institute, 2007).

Both the US and the EU report positively about the numbers of jobs created from biofuels: between 147,000 and 200,000 in the US (Worldwatch Institute, 2007) and 515,000 new European jobs from biomass fuel production are predicted by 2020. If these figures are correct (see comment above), they still have to be set against a loss of jobs in the fossil fuel sector and a shrinking rural population, for example, the number of people directly employed in farming in the US Midwest is a third of the level it was in 1940 (Worldwatch Institute, 2007).

However, what is the 'quality' of the jobs created in biofuels agribusiness? In Brazil, the breakdown is estimated as follows: 30 per cent skilled positions, 10 per cent semi-skilled and 60 per cent unskilled agricultural and industrial work. In the unskilled category, employment is seasonal and low waged. The level of wages for unskilled workers has been extensively criticised and at least in the north-north-east region of Brazil is considered low even 'by developing countries' standards' (Kojima and Johnson, 2005). Yet in São Paulo in the late 1990s, on average, workers in the sugarcane fields were receiving wages that were 80 per cent higher than those of workers holding agricultural jobs related to other crops. On the negative side, the 'high' wage rates paid in sugarcane cutting have been regarded as a driving force for increased mechanisation in the sugarcane industry (Kojima and Johnson, 2005).

The poor labour conditions in the sugarcane estates in Brazil are well documented, in places condemned as inhumane, with people working under conditions which have been likened to a modern form of slavery (Philips, 2007). While such conditions are generally associated with the north-east of Brazil, the wealthier states are also not exempt from examples of exploitation. For example, Amnesty International reports that, in March 2007, attorneys working for the state Ministry for Labour rescued 288 workers from forced labour at six sugarcane plantations in São Paulo State (Amnesty International, 2008). While the Lula government made efforts to improve the working conditions, for example, by complying with International Labour Organization (ILO) standards, they were opposed by other political forces. For example, an inspection team from the Ministry for Labour released over a thousand people working in conditions described as 'analogous to slavery' from a sugar plantation owned by an ethanol producer in Pará State in June 2007. Following the raid, a Senate commission accused the inspectors of exaggerating the workers' poor conditions. As a result, the inspectors' work was briefly suspended (Amnesty International, 2008). Child labour is also reported in Brazil. Again the government is trying to address the issue and there are reports that child labour is decreasing (Friends of the Earth, 2008).

Agricultural estate work throughout Latin America appears to be blighted with poor working conditions and labour relations, and sugarcane, in particular, being most notorious with a history rooted in exploitation, which unfortunately appears to continue today. Contractual employment is standard and workers are frequently migratory. This means that they often live in very poor conditions, have no opportunity to grow their own food and their terms of employment are not as good as salaried employees. Workers have few statutory rights; for example, risk insurance and unionisation are resisted by estate owners (Network for Social Justice and Human Rights, 2007). Wages are paid on the basis of quantities of cane cut. New cane varieties are lighter with higher sugar content. Since cutters are paid by weight of cane cut this means that a larger area of cane has to be cut to yield the same weight as with the old varieties. In the 1980s in Brazil, a cane cutter would earn about US$4.50 a day for cutting around 4 tonnes of cane. In 2007, a cutter would have to cut more than 15 tonnes in a day to earn about US$3.50 (Network for Social Justice and Human Rights, 2007). Payment can also be through credit in the estate store (Christian Aid, 2009).

The nature of the biofuels crop has an influence on the level of job creation. Data on exact numbers are not readily available particularly since some

Table 3.1 Direct jobs per energy sector

	GJ/workday	Jobs/BOE* (× 1000)
Fossil fuels		
Natural gas (US average)	739	0.03
Coal open cast (UK)	397	0.05
Oil (North America average)	316	0.07
Coal deep mined (UK)	249	0.08
Bioenergy		
Wood energy crops, mechanised (EU)	144	0.15
Cane ethanol, mechanised (Brazil)	9.2	2.40
Palm biodiesel, smallholder (Malaysia)	6.0	3.66
Cane ethanol, unmechanised (Brazil)	2.5	8.87
Farm-forestry developing world	1.6	13.86
Jatropha biodiesel, unmechanised	1.2	18.79
Other renewables		
PV** (manufacturing, installing, O&M***)	53.6	0.37
Wind (manufacturing, installing, O&M)	227	0.09
Combined sources (see article)		

Source Biopact, 1 September 2006, http://news.mongabay.com/bioenergy/2006/09/jobs-per-joule-how-much-employment.html

Notes
*   Barrels of Oil Equivalent
** Photovoltaic
*** Operations and Maintenance

crops are not extensively grown commercially, e.g. jatropha. Table 3.1 gives some estimates for different crops. Soy seems to be particularly poor in terms of job creation. Levels of one person in permanent employment per 167–220ha of soy are cited (Cotula *et al.*, 2008). In Brazil, when new sugar and soy plantations are established, for every 100ha the former creates ten jobs while for the latter it is two (Christian Aid, 2009). In Paraguay, 1,000ha of soy can be managed by between one and three people. A significant part of soy production in Paraguay is owned by Brazilian companies that import Brazilian labour who are paid at subsistence rates (Palau *et al.*, 2008). Oilseed crops tend to be the most beneficial in terms of the numbers of jobs created because they are the least mechanised of the biofuel feedstocks, whereas tree crops normally require much less labour than agricultural crops. Therefore, switching from growing non-tree crops to tree crops could be accompanied by a fall in the numbers of workers (Kartha *et al.*, 2005). An analysis for West Kalimantan, Indonesia, concluded that existing smallholder agriculture supported almost 260 times as many livelihoods as plantations that could be used for biofuel production (Renner and McKeown, 2010: 8) Likewise, opening up degraded land to tree-crop production, while potentially having positive environmental impacts on the soil, would not result in a significant stimulus to employment opportunities. Sweet sorghum, a bioethanol crop, is considered to have good potential for local employment (ICRISAT, 2007).

Land issues: institutional aspects

Food security is strongly linked to issues of access to land. However, as well as this very important aspect, ownership, access to and use of land have much broader connotations with historical, political, cultural and spiritual significance. This section deals with these broader issues of land access; issues related to food security are discussed in Chapters 4 and 5.

Government policies to promote biofuels can increase land values. There is certainly evidence from the state of Paraná, Brazil, that shows a correlation between the increase in areas planted with sugarcane between 1995 and 2005 and the accentuated rise in land values in the same period (Rathmann *et al.*, 2010). In some cases this can provide opportunities for farmers to rent or sell land. Unfortunately, farmers do not always make these choices voluntarily and there is concern that where land ownership or entitlement to use is not clear farmers face eviction. The rural poor, particularly those who do not have official title to their land, are potentially vulnerable to being displaced from their livelihood. There are reports in Tanzania that a thousand rice farmers in the Wami Basin are threatened with eviction to make way for sugarcane plantations (Bailey, 2008). At the extreme this exploitation of resources by outsiders disenfranchises people from their land and internal displacement of rural people by biofuels is a potential risk. The UN considers that in West Kalimantan, Indonesia, 5 million indigenous people are at risk of displacement from expansion of palm oil plantations for biofuel production (Bailey, 2008). Pastoralists, shifting cultivators and women are considered to be the most vulnerable social groups. The farming systems of pastoralists and shifting cultivators are viewed negatively by ministries of agriculture and their advisors, being regarded as inefficient and non-viable, hence they fall outside the policy remit (Cotula *et al.*, 2008). Although women make a significant contribution to farming, only 5 per cent of women farmers are registered as owning the land they farm (IUCN, 2007) so decisions about land issues are often taken without their participation (see below for further discussion on gender issues). A woman's land is often registered in the name of a male family member. Widows and single mothers are regarded as particularly vulnerable to losing entitlement to land when decisions are made on their behalf by male family members (DFID, 2007). Women are often unaware of their rights under statutory law (Kameri-Mbote, 2006) and so do not challenge attempts to undermine their rights.

The rate of expansion of large plantations for biofuel crops can be problematic. The expansion of soybean cultivation in Brazil, from 3Mha in 1970 to 18.5Mha in 2003, has displaced many small-scale farmers with meagre compensation. In Santarém in the state of Pará (Brazil), between 2000 and 2003, 600 families sold their land to plantation owners, while in Mato Grosso, the number of farms smaller than 10ha decreased from 23,900 in 1980 to 9,800 in 1996 (Cotula *et al.*, 2008). It is not only the number of families affected but also that communities are being destroyed. For example, in the case of Santarém, up to 70 per cent of the population in some communities were displaced. Although these land changes in Brazil cannot be entirely attributed to crops for biofuels, there is a concern that the changes described are likely to escalate in any dramatic increase in demand for biofuels.

Consolidation of landholdings is in part driven by the need to keep costs down through the mechanisation of biofuel feedstock production, which is better suited to large land holdings. This is not a new phenomenon. The expansion in sugarcane production, occurring in Brazil during the 1970s, was accompanied by large plantation owners taking over smaller-scale farmers' land. Such actions, particularly in north-east Brazil, were sometimes accompanied by violent social disruption (Worldwatch Institute, 2007). Indeed, as biofuel production is increasing so are reports of conflicts. In Brazil, there are renewed reports of intimidation and violence to force people from the land, despite their customary rights to land holding being recognised in statutory law (Van Gelder and Dros, 2006). Unrest is not confined to Brazil. In 2007, about 400 communities in Indonesia were reported to be involved in palm oil-related land conflict linked to the significant expansion of palm oil plantations (Oxfam, 2007). In Colombia there have been unconfirmed allegations of links between paramilitary groups and palm oil companies. There are newspaper reports that paramilitary groups have carried out a 'campaign of killing and intimidation' to drive black and indigenous people off their land to make way for palm oil plantations (Balch and Carroll, 2007).

Land that is not currently being used for agricultural production can be classified by governments as 'degraded', 'unproductive', 'idle', 'marginal' or 'abandoned'. In order to avoid the criticism of crop land being diverted to fuel production, governments are now promoting the use of such land types. However, this is often not how rural people, particularly the poor and more vulnerable, will view this land which can form an important resource for rural households who use it for farming, herding and providing goods and services, such as food, fuelwood, fodder, building materials and medicines (Kartha and Larson, 2000) (see Chapter 4). A survey of villages in Hassan district, Karnataka, India, found that land officially classified by the Ministry of Rural Development as wasteland was viewed very differently by farmers who use the land for grazing livestock and by women who use it as a source of fuelwood, medicines and flowers used in religious ceremonies (Narayanaswamy, 2009). Indeed, crop land was left idle when there were insufficient resources, both human and financial, for cultivation and the farmers' own perception of the land is that it is set aside until circumstances improve when it will be once again be cultivated.

Governments allocate land to developers. This can lead to conflict of land access rights between outsiders (biofuel companies) and local people. The situation arises because of the mechanisms by which both parties consider they derive their legitimacy to determine land use. The government draws its legitimacy from national legislation and statutory law which allows it to grant land rights. Rural people may recognise the formal legal system to legitimise ownership, but the rules they use to determine access to land and the benefits that accrue from that access may be drawn from customary law rather than national legislation. Access to land is shaped 'by social relations, … by relations of power, authority and social identity and by relations of reciprocity, kinship and friendship' (Cotula *et al.*, 2008: 9). Developers, particularly foreign investors, operate within formal legal systems and their actions are guided by what is written in contracts and title deeds. This dichotomy creates the potential for misunderstanding and conflict.

It is quite possible that land not currently under cultivation is remote and hence if export markets are being served, investors might find the quantities of land available are negatively offset by the distance to the nearest port that acts as a deterrent for investment. This appears to have been the case in the United Republic of Tanzania (FAO, 2008a). The procedures for investors acquiring land can be complex and time consuming; for example, in Tanzania, the procedures for leasing land owned by villages can take up to two years (FAO, 2008a).

However, it is not only direct government action that can threaten poor people's access to land, but changes induced by government policy, for example, when promoting biofuels. Land tenure may become more formalised in response to increased land prices as landowners seek to obtain a better return on their assets. Land price increases may well be driven by a better return on agricultural products, both biofuels and food crops. In Brazil, it is reported that some large landowners who had been willing to cooperate with the land reform process, after seeing the higher economic returns beginning to materialise from biofuels, were starting to show signs of reluctance to redistribute land (Cotula *et al.*, 2008).

Box 3.3 Case study: 'Jatropha comes to Kisarawe'

Mtamba, in the coastal district of Kisarawe, is one of 11 villages forming a circle within which Sun Biofuels Tanzania Ltd, a subsidiary of British company Sun Biofuels plc, is about to invest $20m in 8,200ha for jatropha, of which Mtamba owns the majority. Together, the villages are home to about 11,000 people, 850 of whom live in Mtamba.

Although uncultivated, the land is used by the villagers of Mtamba, principally for charcoal making, firewood, and collecting fruits, nuts and herbs. Mtamba was invited to a meeting of all 11 villages with Sun Biofuels to discuss the investment, but their invitation did not arrive until after the meeting had taken place. They were soon visited by the District Land Officer who urged them to make a quick decision, sparking a hastily convened meeting at which the investment was agreed in principle.

However, the first that many of the villagers knew about the scale of the investment was when they saw men laying beacons marking out the area for development. They still do not know how much land they have conceded, but many of the villagers are convinced that this is a big opportunity. 'They're giving us seeds and a market, so this is good for the villagers', says Mussa Mrisho, a local farmer.

Despite the investment being in its final stages, confusion still reigns. According to local press reports, the 11 villages were entitled to total compensation of 800m Tanzanian shillings (about $630,000) – equating to about US$77 per hectare. However, Sun Biofuels has confirmed compensation of $220,000 to be shared between 152 people with trees on their land, and a further $10 per hectare – suggesting total compensation of less than half that reported in the press.

In Mtamba, most do not know whether they will receive any compensation. The Village Council received a letter from the District Land Officer requesting villagers to apply for compensation. But the village committee was unsure what to do. As a result, they say only six people have returned it. The deadline has now passed. The District Land Office says that everyone who is receiving compensation has been informed.

Although they do not know how much land they are actually conceding to Sun Biofuels, the villagers do know it includes a waterhole which is the only place where they can collect water during dry periods. They also collect clay there to build houses. They say they have had assurances from Sun Biofuels that they will retain access to the waterhole and clay once the development is under way. However, they have nothing agreed in writing and when asked about this, Sun Biofuels was unaware of the waterhole.

What the people of Mtamba really want are jobs. During a meeting with Sun Biofuels, they were told that 4,000 of the 11,000 villagers in the area would be employed. Two hundred people from Mtamba have applied for jobs as drivers, guards and farmers, but none have heard anything back. Sun Biofuels estimate that there will initially be about 1,500 jobs to clear the land and in the longer term expects to create one job for each hectare. The villagers have been told that they will be given priority, but they remain uncertain, and wish they had something in writing to confirm this.

Source: Oxfam research, including interviews, field visits and desk research (Bailey, 2008)

What does the case study from Tanzania (see Box 3.3) show? First, poor communication between the parties can later lead to problems when villagers realise the rights they have conceded. Second, unclear rules and processes, particularly for rural people who are not always confident about dealing with officialdom, do not lead to equitable outcomes. In Indonesia, some communities consider that agreements they have reached with palm oil companies are only for temporary transfer of land-use rights whereas the companies consider that they have the permanent transfer of rights (Cotula *et al.*, 2008). These sorts of negotiations are not taking place between equals, and there is concern that communities are being offered basic development and services, such as roads and schools, only when they give consent to land being used by project developers rather than receiving these facilities under transparent planning criteria (Freeman *et al.*, 2008). In some cases there are concerns that they are not consulted at all and are only informed of decisions.

Since the company in Box 3.3 has its roots in Europe, it may respond to civil society pressure to ensure that it meets the expectations of local people, such as access to the waterhole. However, other trade agreements stand to undermine government's options to regulate foreign companies' behaviour. For example, it is possible that any attempts to restrict biofuel imports using sustainability criteria based on labour rights and food security could contravene the General Agreement on Tariffs and Trade (GATT) (UNCTAD, 2008).

There are weaknesses in the institutional framework in many developing countries. As a consequence poor people are vulnerable to exploitation by outsiders. Planning processes can be abused and bypassed unless there is strong support for enforcing rural people's rights by the government. For example in Mozambique, communities that had already been displaced from their traditional lands by the establishment of a conservation area, were facing eviction from their new site to make way for a 30,000ha sugarcane plantation planned by a foreign investor on land allocated by the government (Cotula *et al.*, 2008). In 2003 in Mato Grosso, Brazil, large-scale farmers were found to be illegally occupying 3.2Mha of public land (Cotula *et al.*, 2008). In Colombia, palm oil companies have been found to be illegally occupying land which had been allocated by the government to displaced indigenous communities in Curvaradó and Jiguamiandó. These communities have had to engage in a lengthy legal process to regain their land (Cotula *et al.*, 2008). Palm oil production in Indonesia has become tainted with reports of companies not keeping to agreements about land acquisition. Community members can donate land to a palm oil scheme. The large-scale producer keeps a portion of the land and the community members receive back a small allocation. However, in practice, the allocation can be smaller than had been agreed (Cotula *et al.*, 2008).

All too often, laws can be formulated in such a way that they can be exploited in favour of business interests rather than those of rural people. In Indonesia, customary land rights are recognised by statutory law, although all land without a registered title deed is regarded as state land which puts the decision-making power over who uses the land in the hands of the government. The law allows people to use non-registered land as long as the exercising of these rights is consistent with national interest. Of course, the interpretation of vague terms, such as 'national interest', is likely to be contentious particularly since it puts a lot of influence into the hands of government agencies with wide discretionary powers and weak countervailing powers creating a system with potential for abuse. Indeed, in Indonesia, it appears that land used under customary law can be 'taken for a public purpose, which includes business activities run by private corporations (article 18 of the Basic Agrarian Law and subsequent instruments)' (Cotula *et al.*, 2008: 41).

Even when a scheme appears to be an example of good practice in theory, translation into practice can lead to outcomes different than intended. Mozambique has a mandatory requirement for community consultation which is a first step towards the negotiation of a benefit-sharing scheme between communities and investors requiring use of land. Based on experiences not specifically related to biofuels projects, it would appear that consultations do not always take place (at least there are no records) and sharing benefits is limited to very small one-off payments, rather than equitable distribution of long-term benefits. Consultations that have taken place have tended to involve only a few community members, generally local elites who are able to monopolise the benefits. A particular concern is the lack of any effective monitoring scheme to ensure investor compliance with any agreements (Cotula *et al.*, 2008).

There can also be conflict between statutory and customary law. Usufructuary rights can be granted to plant biofuel crops. However, the type of crop can cause conflict. For example, in Mali, under customary law planting a tree can confer ownership of land to the person who plants the tree (Practical Action Consulting, 2009). This simple act of planting a tree can open a Pandora's Box for conflict. Tree planting by tenants will be resisted by landowners, while tenants might see it as an opportunity to acquire land. Alternatively, tenants might only be willing to invest in short rotation crops. It has been reported that once degraded land, which has been considered as common land, has been designated for biofuel crops, the owners suddenly make themselves known to the authorities in order to claim benefits (ICRISAT, 2007).

However, the displacement of poor farmers and the injustices faced by communities is not an inherent characteristic of biofuels but they are a consequence of asymmetries in power relations in a particular context. Large-scale commercial biofuel projects typically involve a range of actors, including project developers, investors, government agencies and local communities with very different negotiating powers, skills and resources. These actors exhibit differences in their capacity to influence decision makers and opinion formers, to mobilise political support and to draw power from parallel processes of negotiation (Cotula *et al.*, 2008). Developers and investors in biofuels are often major international companies involved in the energy or agriculture sectors. Such companies are able to command huge financial resources and can hire or have in-house highly competent legal teams whereas communities have very limited resources and often have to rely on civil society organisations to provide the skills and finance that they may need to protect their interests. In other words, such actors dominate the governance of biofuel value chains. Communities are also not a homogeneous group and divisions can arise along lines of wealth, gender, age and social status. Local elites, such as customary chiefs, may enter into agreements with the government and biofuel investors that are contrary to the views of other community members. An example comes from palm oil expansion in Sanggau district, West Kalimantan, where some Dayak community leaders are reported to have aligned themselves with a palm oil company operating in their district, in return for personal benefits such as priority access to smallholdings (Colchester *et al.*, 2006).

Gender issues

Gender issues in biofuels programmes emerge in terms of access and control over household assets and in terms of income-generating opportunities for women within biofuels programmes, either as growers of biofuel feedstocks or as employees in a biofuels agribusiness. Given the rather limited number of biofuel programmes there is little direct empirical evidence about gender issues specifically related to biofuels. Indeed, women's representatives at the 2007 session of the UN Commission on Sustainable Development, which included discussions on energy for sustainable development, urged governments to document best practices in biofuel production that adopted gender-sensitive approaches (Karlsson and Banda, 2009). The literature available at the time of writing tends to draw on the experiences of women and men in general agricultural production for export markets and by extension drawing conclusions about the likely effects of biofuel production rather than from direct experience with biofuels.

Poverty has a distinct gender dimension in the sense that poverty is experienced differently by women and men and their responses are different (Sen, 2008). Women and men have different degrees of access and control over land and natural resources (Rocheleau *et al.*, 1996). They also have different roles, responsibilities, rights and obligations which define the relationship between women and men in a household. This relationship is defined and bounded by the power to make decisions about, and exercise control over, a way of living and resources. In most societies power generally resides with men. At the household level in rural areas, gender relations result in women usually being excluded from decision making about which crops can be planted and where they can be planted. This lack of control also extends to decisions about rehabilitating and restoring land; possibly problematic for women when land classified as 'wasteland' is designated as appropriate for biofuels. This exclusion from decision making is closely linked to social institutions that prohibit women from owning land (Kelkar, 1995). This means that women might be excluded from a decision about whether or not to grow biofuel feedstocks on land owned by the household, the consequences of which could undermine household food security. It has been argued that the income and employment benefits of producing cash crops are not spread equally within households (von Braun and Kennedy, 1994; Gladwin and Thomson, 1997; Govereh and Jayne, 1999; Wandschneider and Garrido-Mirapeix, 1999; Murwira, Wedgewood *et al.*, 2000). Once agricultural production becomes commercialised, the role of women as food producers is marginalised because men take over and assume control over household assets.

Women's levels of land ownership are considerably less than men's; for example, in Cameroon women own 10 per cent of the land, in Brazil 11 per cent and in Peru 13 per cent (Rossi and Lambrou, 2008: 5). Where women do exercise control over land, they face barriers to participation in new initiatives requiring significant levels of investment, since they have difficulties accessing credit because they often lack title to land, livestock or other property required as collateral for loans. A requirement for entering into a financial contract as a feedstock supplier can also exclude women in some cultures where they are regarded as legal minors.

Women are regarded as being particularly vulnerable when common land (which could include 'wasteland') is diverted to biofuels feedstock production. Because of their limited control over private resources, women draw on common property resources for many goods and services for meeting household needs, such as food, fuel, building materials and medicines (see for example Rocheleau *et al.*, 1996; Karlsson and Banda, 2009; Narayanaswamy, 2009). In northern India, it is estimated that nearly half of the income of poor women depends on resources from common land compared to only one-eighth of poor men's incomes (Reddy *et al.*, 1997). The poorer the household, the higher the contribution common land makes to meeting household needs (Gundimeda, 2005, quoted in Rossi and Lambrou, 2008: 6). Access to common property resources can be threatened when traditional use rights are disregarded and overridden by official legal agreements (Kartha and Larson, 2000).

There are examples of biofuels projects that have taken gender issues into consideration, and where women as well as men have benefited. For example, a project to develop a model for decentralised electricity generation based on jatropha has been established in a remote village in Chhattishgarh, India, which aims to ensure gender equity in both the governance and access to income. A specially established village energy committee (VEC) oversees seed collection and is responsible for collecting the payment for electricity bills. The VEC is elected by the villagers with women among the representatives. Women's self-help groups are able to earn an income from seed collection (Practical Action Consulting, 2009).

However, women do appear to be prepared to participate on more adverse terms of incorporation than men. Men do not participate when the rates of remuneration are considered too low. For example, in Zimbabwe women are participating in jatropha cultivation while men are excluding themselves. The women value the access to the oil and by-products whereas the men consider growing maize likely to yield a higher return than jatropha (Karlsson and Banda, 2009). On the other hand, the women find that jatropha, at least in the early stages of its establishment, is labour intensive and harvesting can compete with maize and other food crops, which they consider a priority (Karlsson and Banda, 2009).

In Brazil women are opting for exclusion from the biodiesel programme (the PNPB) and are organising resistance against inclusion in the programme. In Maranhão women control the harvest of babaçu palm seeds and the sale of the oil production. The women consider that the government's programme is at a scale incompatible with their present level of extraction and production. They fear that the economic inclusion of babaçu production into the national biodiesel programme will cause many households to become involved in land conflicts that will be lost by women because of their lack of ownership of land titles. As a consequence their carefully built economic power (in the household sphere) and political power (in the wider society) will be weakened or destroyed (Hospes and Clancy, 2011).

Women have also benefited indirectly when biodiesel has been used for decentralised electricity generation. Women in a remote district of Cambodia have switched from pedal sewing machines to electric ones when an entrepreneur in their village produced electricity using a biofuel diesel generator. Such a transition increases their output and hence, if there is sufficient market opportunities, increase their income (Karlsson and Banda, 2009).

Does biofuel production create employment opportunities that enable poor women to move out of poverty? There are two particularly relevant issues here. The first is poor women's ability to participate in paid labour markets, and second is the level of remuneration that they would receive in those markets. In terms of the former, there are questions about whether or not poor women have the time to participate in labour markets (Sen, 2008).

If poor women are earning good levels of remuneration by participating in biofuel markets it would be significant (and surprising) since all labour markets are characterised by unequal opportunities based on gender, with women concentrated in lower quality poorly paid employment (Heintz, 2008). Given that the extent of biofuel production is still limited, the literature on gender disaggregated data employment is deficient and limited to Brazil. A study by FAO found that women participated in the cane fields along with men. Women formed about 15 to 20 per cent of the total workforce with some seasonal variations (Rossini and Alves Calió, undated). The ethanol programme of the 1970s has been credited with enabling a 'large number' of women[3] to enter the labour force. In part this transformation can be attributed to the loss of smallholdings that had formed the basis of the women's livelihoods, and in part that new opportunities were perceived as routes out of poverty. Although the FAO study could not determine the exact remuneration rates, women were considered to earn less than men since their productivity was classified as lower. Mechanisation, although potentially creating the possibilities for women to do the same tasks as men, seems to have become a male preserve (Rossi and Lambrou, 2008). There are reports that the numbers of women cane cutters are falling (Friends of the Earth, 2008). Most worryingly the Andradina Rural Workers Union,

(Ser Andradina) reports that some plantations are asking women to show proof of infertility before they are employed, presumably to avoid payment during maternity leave (Friends of the Earth, 2008). In addition, the wages of women cutters in Costa Rica are reported to be paid to their husbands or male partners (Network for Social Justice and Human Rights, 2007).

Rossi and Lambrou (2008), drawing on experiences from agriculture in general, warn that the prospects for women as paid employees in biofuel crop production are not positive and are potentially not as good as men's prospects. Women are increasingly employed in agriculture and they now constitute about 20 to 30 per cent of the waged workforce with a tendency to seasonal, causal or temporary work (Rossi and Lambrou, 2008: 14). Such employment terms usually result in limited benefits, including medical treatment, and a tendency for less training than men. Oxfam has expressed concern that in the oil palm plantations of Malaysia women are recruited as sprayers of chemical pesticides and herbicides often without proper training and equipment, which may have long-term health consequences (Oxfam, 2007). However, the application of a gender-sensitive approach can help women improve their skills levels and hence income, as well as possibly their status within the family. In a Biofuel Africa project in Ghana women have been trained as tractor drivers which has resulted in them becoming the main income earner for their family, with a monthly income of 150 cedi (approximately €77) (Baxter, 2010).

Conclusions

At the time of writing there is limited empirical evidence about the social impacts related specifically to biofuel production. Most of the concerns are based on experiences with plantation agriculture in general. However, there are indications that, for small-scale farmers, growing biofuels offer opportunities for product diversification and will enable them to compete with farmers producing on a larger-scale. The challenge is to ensure that small-scale farmers get a fair deal. A number of ways for small-scale farmers to benefit are beginning to emerge as well as different types of contracts which can offer options to be matched to local realities. Farmers also need support when growing new crops.

Biofuels are not pro-poor where rural people do not have clear title to their land. There is a growing body of evidence which catalogues injustices to rural people linked to biofuels expansion by large-scale agribusiness. These injustices are rooted in power asymmetries in which large companies, with or without collusion by government officials, are able to gain control of the resources for biofuel production. It should be stressed that this is nothing that is inherent in biofuels but is related to the power struggles linked to large-scale cash crop production. The threats to land availability are not only from crops for biofuels, but also the phenomenon of wealthy countries with poor or limited agricultural soil, that to ensure food security, are buying up large tracts of land in another country to grow their staple grains (Haralambous *et al.*, 2009). Also, the social problems in the sugarcane fields of north-east Brazil and the palm oil plantations of Indonesia existed before biofuels. Of course fears that the social situation will deteriorate as a consequence of biofuels expansion are justified in some contexts and should be resisted. There are signs of resistance, for example the Via Campesina, an international movement of poor peasants and small-scale farmers in the developing and industrialised worlds, which campaigns against biofuels and other agribusiness (Borras *et al.*, 2010). However, biofuel production does not automatically result in social deterioration in rural areas. There is beginning to emerge in the literature reports of biofuels being produced using socially sustainable methods.

The opportunities for landless people who have to sell their labour to estate grown biofuels are limited as are the number and quality of jobs available for unskilled workers. The major cost component in biofuel production is the feedstock. Therefore, where a choice of biofuel crops can be made, tree crops might be the preferred option since they have a generally lower labour demand than annual field crops. Where there are existing biofuel feedstock crops, such as sugarcane, there will be a move towards mechanisation to reduce labour inputs, which would disadvantage the unskilled workers. While the working conditions in the cane fields can be atrocious, landless rural people need employment. It is the role of the state to ensure that employment laws are enacted and enforced to provide better working conditions. International standards can contribute to improvements in working conditions (see Chapter 7).

While rural women suffer the same as rural men from the negative effects of biofuels described in the last two paragraphs, the other specific dangers identified for women, such as family land being diverted to biofuels and so undermining household food security, do not as yet appear to have materialised or at least have not been reported. There are signs that women have actually benefited directly from biofuels projects, usually in small-scale developments involving non-governmental organisations (NGOs), and indirectly where biofuels are used for improving local energy services. In terms of plantation employment, there is evidence to suggest that the unique needs of women workers, e.g. related to child care, are neglected.

Therefore, in conclusion we can say that the evidence on the social impacts of large-scale biofuels is mixed. Where no attempt is made to address asymmetries of power, then biofuels are not pro-poor and there is resistance to inclusion. However, where conscious efforts are made to ensure equity of participation and equity in distribution of benefits, the rural poor can benefit and they actively welcome the opportunity for diversification in income generation.

4 Biofuel production and ecosystem services

Introduction

Growing biofuels, as with any agricultural activity, affects the natural environment with impacts on habitats which form the earth's myriad of ecosystems.[1] The type and extent of the effects depend on the biofuel crop, the agricultural practices employed to grow it and where it is grown. There is considerable concern when virgin ecosystems are destroyed to plant biofuels but there are also ecosystem changes when other types of land, such as agricultural land and non-productive land are converted to biofuels. A substantial change to natural ecosystems alters the flow of 'goods and services' (collectively commonly known as *ecosystem services*)[2] emanating from them. There is no standard definition of 'ecosystem services' but they can be considered to be 'the benefits of nature to households, communities and economies' (Boyd and Banzhaf, 2007). The goods can be of direct value, such as provision of grazing, timber and other products; or other forms of socially and economically important services such as carbon sequestration or maintenance of the hydrological cycle for water provision or flood amelioration (Ash and Jenkins, 2007; TEEB, 2010).

Ecosystem services are constantly under threat from a range of development activities not only biofuels. However, what is a cause for concern is the rapid expansion of land devoted to biofuels and ecosystem services. In 2007 biofuel crops used about 1.7 per cent of global cropland (26.6Mha) with increase in production to 2.3 per cent of cropland (35.7Mha) in 2008 (UNEP, 2009). Projected increases vary enormously depending on the type of feedstock, where it is grown and possible yields with conservative estimates being about 5–10 per cent of global cropland being devoted to biofuel crops by 2020 (UNEP, 2009). The negative reactions to the use of agricultural land for growing fuels have resulted in other land types being considered for production. In some cases, such as Indonesia, natural forests have been cleared with the intention of planting oil palm. Such actions destroy the ecosystems and the services they provide which have added to the negative public reaction to biofuels. As a consequence other options are now being promoted as possible ways of increasing outputs by extending the land area under biofuel crop cultivation, such as bringing degraded land back into production, increasing productivity on land currently farmed with low input extensive systems and using 'marginal' lands (Hart Energy Consulting and CABI, 2010). However, 'degraded land' is a value laden term and, from an ecological perspective, all land types have an ecosystem that is capable of providing ecosystem services.

In this chapter we examine whether or not it is possible to reconcile two interconnected problems: can biofuel production be 'biodiversity friendly' and, if so, can production be achieved in a pro-poor way in the sense that access to ecosystem services by low income households is sustained. So here we differ from Chapter 3 in which biofuels were examined from the perspective of how smallholders could benefit from direct participation in producing biofuels, whereas in this chapter we look at how biofuels indirectly affect household services and income through land-use changes as a consequence of biofuel production. The ecological impacts of large-scale commercial biofuel production in agricultural plantations centre on three main issues: transformation of natural vegetation, impact on soils including the carbon content with the consequences for greenhouse gas storage or release, and water use. In this chapter we deal with the first and third issue, while carbon content of natural vegetation was discussed in Chapter 1. As was explained in Chapter 1, we do not address in detail the impacts of climate change on the poor not because we are climate sceptics but because the focus of this book is the direct and more immediate impact of biofuels on rural poverty, whereas the climate change impacts related to biofuels are indirect.

This chapter is divided into five sections. The first describes ecosystems and their services and their relevance for the rural poor. The second explains the relevance of biodiversity to ecosystem sustainability, while section three presents an overview of the impact of biofuel plantations on ecosystems. The fourth discusses the impacts of biofuel plantations on a major ecosystem service: water resources. The chapter concludes with a discussion as to whether or not biofuel production can both be 'biodiversity friendly' and not undermine the security of low income households in rural areas.

Socio-ecosystems and their services

An ecosystem is a habitat that supports a particular set of species of plants, animals and micro-organisms. Terrestrial ecosystems are often classified according to their vegetative cover: lowland forest, dry deciduous forest, grassland or wetlands. Aquatic ecosystems are classified as marine (subdivided by location, such as open ocean, sea bed and coastal) or fresh water (which can be still, such as ponds, or flowing, such as rivers). However, it is possible to sub-divide these further for a more detailed analysis (Millennium Ecosystem Assessment, 2003). Indeed there can be much discussion about where to draw the boundaries of an ecosystem, and so they vary in scale (global, regional and local). It is even possible to conceive of the planet as one large ecosystem which can be divided into numerous subsystems.[3]

The living things within an ecosystem interact with each other and their physical environment in complex ways that we are only just beginning to understand.[4] The systems are dynamic: their characteristics, resilience and stability change over time as social systems and climate patterns and other influencing factors change. Changes in the ecosystems can lead to a greater abundance of pests and diseases that affect humans, plants and animals. Ecosystems are connected and there are flows of material between them. We have limited understanding of these processes, as well as the way in which social systems interact and mediate them, in part because the ecological and socio-economic systems are usually studied separately, since they are part of different discourses.[5]

The social system interacts with the natural system through people's dependence on ecosystems for a wide range of goods, e.g. food, fuel, water, timber and medicines; and services, e.g. flood control and climate regulation; as well as endowing the natural world with cultural significance where the natural landscape, e.g. forests and springs, can be places of spiritual or aesthetic value (Ash and Jenkins, 2007; TEEB, 2010). At the local level ecosystems provide many people with the basic necessities of life as well as providing sources of income. In other words ecosystems are not only instrumental for human well-being (Daily, 1997) but can also be considered as constituent elements of well-being (Millennium Ecosystem Assessment, 2003). For most people, particularly those living in rural areas, the way they value the natural world is through the services it provides rather than in terms of remote scientific concepts of biodiversity. However, as we describe in the next section, it is biodiversity that underpins many of these services.

Measures of the extent of the human use of ecosystem services are difficult to obtain and there is little available data on the quantities and values of these services. Yet, it is reported that in South Africa 94 per cent of canopy and 77 per cent of sub-canopy forest species have at least one recorded use (Geldenhuys, 1999, cited in Shackleton *et al.*, 2008: 77). Information is fragmented, tending to look at one service and in one locality rather than in a holistic manner (Shackleton *et al.*, 2008). Indeed, the contribution of ecosystem services to household income is generally neglected in national income accounts, so impacts that lead to the degradation of ecosystems are also neglected by policy makers (World Bank, 2008a). While it is largely recognised that the poor are the main users of ecosystem services at the community level, wealthier households also draw on the services. However, the information available does not contextualise the findings, in particular about the other options open to the poor for providing their needed goods and services.

People are also part of ecological systems and they are strong drivers of change in the structure or function of these systems. The relationship between people, ecosystems and the derived services are complex and vary over space and time. Figure 4.1 shows an attempt by the Millennium Ecosystem Assessment (MEA)[6] to map out these relationships in which ecosystem services are linked to four groups of criteria said to describe human well-being. Human intervention can increase some services, in terms of quality and quantity, although this can be at the expense of other people's services or other types of services. The effects of these interventions vary in time, scale, space and intensity. The interventions can lead to a complete change in land use, for example, when woodland is cleared for agriculture. The drivers, or factors, of change are complex and contextual. They include population growth, conflict leading to migration of large numbers of people, increased household wealth leading to increased animal ownership, decrease in crop

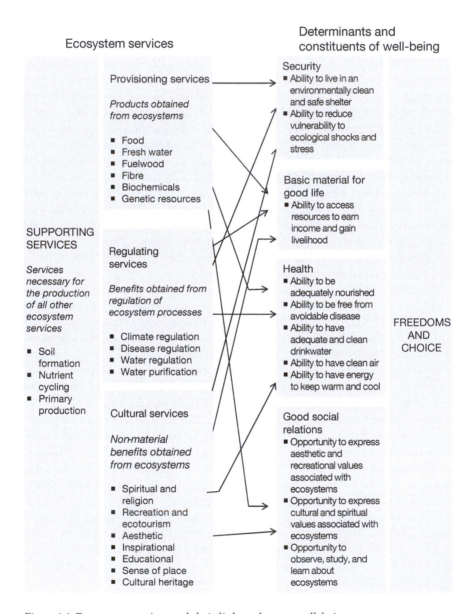

Figure 4.1 Ecosystem services and their links to human well-being

Source: adapted from Ecosystems and Human Well-being (2003) A Report of the Conceptual Framework Working Group, of the MEA

yields, breakdown of informal regulating institutions and increased participation in the informal sector through selling natural products, such as fuelwood and charcoal.

The anthropogenic alteration of ecosystems is not a new phenomenon but it is the rate of change in the last few decades that has accelerated and increased concern (MEA, 2003). The MEA found that 15 out of the 24 ecosystems they evaluated had been degraded (Haines-Young and Potschin, 2010). The dynamics between social drivers that lead to environmental change brought about by human activities, and the way that the environment responds and in turn impacts on human activities are complex. The changes vary in time, area of impact and intensity. The effects on ecosystem services can be immediate, for example, clearing forest for agriculture, or they can take time to appear, for example, soil salination from groundwater irrigation. Use of an ecosystem service in one location can have a negative impact in another, for example, building a dam for hydroelectric power generation can change water flow patterns downstream affecting downstream communities' use of water ecosystem services.

The effects of changes to ecosystems on human well-being can be either direct, such as dumping waste in streams which poisons fish that people depend on as a source of food and livelihood, or indirect, such as building a dam which creates ideal breeding conditions for malarial mosquitos. The negative impacts are not equally distributed. It is generally the poor who are most vulnerable to ecosystem degradation since they draw on ecosystems more than the non-poor for basic services and as a basis for livelihoods, for example, selling forest products. Ecosystem services are available at no financial cost to the user which means that poor households can use their limited cash resources for other needs, e.g. school fees. A reduction in ecosystem services increases people's vulnerability by reducing livelihood opportunities and by increasing the demands on scarce cash resources. There is concern that the MDGs will not be reached if ecosystems continue to be degraded (Fischlin *et al.*, 2007).

Agriculture creates a highly modified ecosystem that provides people with food as well as fuel and fibre. Each type of agricultural system, for example, coffee plantations, cereal farming and cattle rearing, will have its own ecosystem. These ecosystems can suffer from degradation resulting in a reduction of ecosystem services through soil erosion, salination, compaction and nutrient depletion. Management of this ecosystem by the use of fertilisers tries to ensure there is no loss of ecosystem services either in terms of quantity or quality. While fertility and food output levels are maintained, there may still be soil erosion which will lead eventually to food output levels declining. However, the fertilisers can enter waterways and lead to pollution which damages the services provided by that ecosystem, and may be felt by other downstream communities.

If an ecosystem service is damaged, it might be possible to substitute the service with technology. However, the cost of this substitution might be high which can put the service out of the reach of poor people. For example, if water has become polluted then water treatment plants can provide drinking water. However, this technological solution can be expensive, putting the drinking water out of reach of low income households.

The social system is also affected when ecosystems are changed with some groups benefiting more than others. Socio-ecosystems have complex feedback loops and responses, which are not fully understood. The cultural values and belief systems of many communities are embedded in their natural environment and govern the way people behave towards living things and natural features such as streams and lakes. Therefore, when an area is imbued with cultural and religious significance, it can be left free from human disturbance, facilitating a rich and varied biodiversity (Shackleton *et al.*, 2008). Cultural rituals and practices are considered critical in building social cohesion contributing to a 'sense of community or belonging' (Shackleton *et al.*, 2008). Destroying the natural features of cultural significance can contribute to a breakdown of social cohesion and a weakening of the informal institutions, which have protected habitats and their inherent biodiversity.

Biodiversity and its role in ecosystems services

In the previous section it was stated that biodiversity is the factor underpinning ecosystem services. Biodiversity is a scientific concept used to describe the natural world and the interaction between living things and their environment. Biodiversity is defined by the United Nations Convention on Biological Diversity (UNCBD, 1973) and the MEA (MEA, 2003) as: 'the variability among living organisms from all sources including, inter alia, terrestrial, marine and other aquatic ecosystems and the ecological complexes of which they are part; this includes diversity within species, between species and of ecosystems'.

Biodiversity, in the form of plants and animals, provides multiple sources of ecosystem services of food, fuel and fibre which people rely on regularly or in times of need (World Bank, 2008a). There are less visible but nevertheless essential services provided by microbes, which transform waste and contribute to nutrient recycling and soil fertility, and micro-fauna, which contribute to water quality.

From the above definition it can be deduced that both natural and managed ecosystems exhibit their own biodiversity. However, we only have a limited understanding of biodiversity and there is an imbalance in spatial distribution of scientific knowledge about biodiversity, which tends to be skewed towards temperate climates rather than tropical ones. There is a lack of empirical evidence about the role of biodiversity for maintaining ecosystem services (Mertz *et al.*, 2007). As a consequence, there is no consensus among ecologists about, first, the need for high levels of biodiversity to provide ecosystem services and, second, about the extent and form of reduction in biodiversity that would maintain a specific service.

Nevertheless, even with limited knowledge, it is possible to identify system characteristics where significant change to an ecosystem, such as by converting a natural area to intensive agriculture, can impact negatively on biodiversity composition and levels and hence on ecosystem services. Biodiversity changes in response to external factors, e.g. fire or drought. A resilient ecosystem will over time return to its original state. Resilience appears to depend on the type of ecosystem that is a function of abiotic (non-living) and biotic (living) characteristics including geomorphology, climate and biodiversity (MEA, 2003). Ecosystems with high biodiversity, that is when they contain many species, are generally considered better able to cope with stress, that is they are more resilient to major changes in environmental conditions such as drought, than ecosystems with low biodiversity (MEA, 2003). Species loss in dryland systems appears to reduce resilience more quickly than in more humid systems (Madzwamuse *et al.*, 2007, cited in Shackleton *et al.*, 2008: 81). However, there are factors that have such an effect on an ecosystem that it does not recover and a new system emerges, usually with reduced levels of biodiversity and, by extension, a reduced level and/or different ecosystem services.

It is not only important to maintain species biodiversity, but also the gene pool in wild species is critical for ensuring our adaptation to changing environmental circumstances, such as climate change. This is one of the reasons for objections to genetically modified organisms (GMOs) being introduced into farming systems, including, possibly, for the production of biofuels, since we do not know how they will affect this wild gene pool, nor the long-term effects of eating wild plants that have acquired a modified gene on human and animal health. This reduction in the gene pool, therefore, threatens species' resilience to changes in their environment such as drought and, in the longer term, climate change (Shackleton *et al.*, 2008). The pressure to use GMOs for agricultural production comes from the need to increase yields without land expansion and without more intensive use of inputs, such as pesticides and fertilisers (Sexton *et al.*, 2009). For first generation biofuel crops, genetic modification that increases yields per hectare could possibly reduce negative impacts on food production by reducing the pressure for good quality land. There are commercially available genetically modified versions of two common biofuel crops: maize and soy. Crops which have been genetically modified are supposed to be grown under strictly monitored conditions. However, there are reports of an illegal trade in genetically modified soy in Brazil, which means monitoring systems are being evaded (Bickel and Dros, 2003).

Anthropogenic change of clearing natural forest for agriculture can have significant impacts on biodiversity. Replacing species-rich native vegetation, such as tropical rainforest, with an agricultural cropping system based on a relatively small number of species will reduce biodiversity. Even tree crops, such as palm oil, generally contain significantly fewer species than native forest. Compared to natural forest in the same regions, the richness of vertebrate species in oil palm plantations was 38 per cent that of forest, with only 23 per cent of forest species occurring in plantations; plantations also lacked 'iconic' species such as tigers and clouded leopards (Danielsen *et al.*, 2009).

However, invertebrate species richness was comparable between forests and plantations and in some cases it was higher (Danielsen *et al.*, 2009). Palm oil plantations were also found to affect biodiversity in adjacent habitats through fragmentation, edge effects and pollution. On the other hand, palm oil plantations, since they are perennials, have more species-rich ecosystems than systems based on short-term rotation crops.

Species-rich vegetation is not evenly spread over the planet, so in the context of the ecological implications of the establishment of biofuels plantations it is relevant where they are located. For example, vegetation in most temperate regions, such as much of central and northern Europe, contains far fewer and more widespread species than formations such as tropical rainforest. So in terms of biodiversity loss, transformation of tropical forests into biofuel plantations is likely to have a greater effect than biofuel production in temperate regions.

High biodiversity levels are also associated with more productive systems in terms of the quantity and range of ecosystem services. Therefore, it can be reasonably concluded that maintaining species biodiversity is important for the poor, because it meets not only current needs but also allows for future livelihood opportunities based on species currently not utilised, for example, for medicines. It is not only the quantity of a particular species available, but also the quality (such as the size of a fish or tree) that is important for the poor. However, people outside of the scientific community are often more concerned about the abundance of particular species, for example, to provide food; or are pests; or iconic species, such as tigers, rather than having many different species.

Impacts of biofuel feedstock production on ecosystems and their services

Biofuels can lead to direct land-use changes (LUC) when natural ecosystems are cleared for planting biofuel crops or existing agricultural crops or pasture are displaced. Agriculture is considered the most significant driver of ecosystem modification (MEA, 2005). Between 1980 and 2000 agricultural land area in developing countries increased by 629Mha; in contrast, developed countries lost 335Mha (Gibbs *et al.*, 2010). During this period more than half (55 per cent) of new agricultural land was derived from intact forests with nearly a further third (28 per cent) from disturbed forests (Gibbs *et al.*, 2010). A large proportion of this expansion is estimated for crops that could *potentially* be used for biofuel feedstocks. Since the feedstock can be switched between chains, it is difficult to estimate how much of this cultivation of new land can be attributed to biofuels. As of 2010, Brazil had 21Mha of soy beans (compared to 13Mha in 2000) and oil palm in Indonesia increased from 2Mha in 2000 to 5Mha in 2008 (reviewed in Gibbs *et al.*, 2010).

There is a concern that when biofuels substitute for an existing crop, production of the latter does not always cease but is displaced elsewhere. The displacement can be within the same locale, another region in the same country or in another country. Linking these indirect land-use changes (ILUC) to biofuels is difficult to prove, particularly in view of the fact that many crops that can be first generation biofuel feedstocks are also part of food chains producing for both human and animal consumption, and growers are able to switch between chains.

Increased biofuel production can result in the transformation of vegetation consisting of diverse native species, with its associated animal and micro-organisms, into intensive agriculture. Intensive agriculture is dominated by a relatively small number of economically desirable crop plants, which can be exotic in origin, i.e. not native to the ecosystem, region or country. Over time, agricultural systems evolve their own ecosystem, although with biodiversity levels that are less species-rich than under the previous natural ecosystem. These land-use changes, particularly the use of exotic species, raise concern for the conservation of native species. The land type which gives rise to most public concern is the species-rich tropical rainforests, which contain around 70 per cent of the planet's plant and animal species (Donald, 2004). However, other equally sensitive ecological areas considered under threat from biofuels receive less public attention: savannahs and grass lands, indigenous forests, peat and wetlands.

As was pointed out above, biodiversity is the basis of many ecosystem services. Therefore, reduction in biodiversity is likely to be accompanied by a decline in ecosystem services. Any land-use change resulting in a monoculture will lead to a significant loss of biodiversity. Table 4.1 gives an overview of the potential impact on biodiversity of the most common biofuel feedstocks. The impacts of biofuels on biodiversity are considered to arise from habitat destruction, invasive alien species, pollution and climate change of which habitat destruction is considered to be the greatest threat. However, it appears that only the impacts of palm oil on biodiversity have been studied extensively.

The areas where biodiversity is considered most at threat are the tropical regions of South East Asia, from palm oil, and Latin America, from soy (Stromberg *et al.*, 2010). Soy is considered to be a significant threat to

Table 4.1 Feedstock specific potential impacts on biodiversity

Feedstock	Land usually converted or used	Impact on biodiversity
Oil palm	Virgin forest	Very high
Sugarcane	Grassland/cultivated land	High
Maize	Cultivated land	High
Cassava	Cultivated land/grassland/forest	Neutral
Rapeseed	Grassland	High
Soybean	Cultivated land/grassland/forest	Very high
Jatropha	Grassland/cultivated land	Neutral

Source: Stromberg *et al.*, 2010: 27

biodiversity, both from clearance of natural habitats to plant the bean and from clearance for the supporting infrastructure, such as roads and railways, which further open up access to land (Donald, 2004). An area thought particularly vulnerable to biofuels expansion, particularly sugarcane, is the *cerrado*, an area of wild savannah which covers more than 25 per cent of Brazil's land mass. Half of Brazil's endemic species (that is, species found nowhere else on earth) and a quarter of its threatened species are found on the *cerrado* (Worldwatch Institute, 2007). Soybean cultivation in Brazil has brought land-use changes as a consequence of large-scale farmers acquiring small plots of land which were used to grow a wider variety of crops but now support a monoculture. While the soybeans have primarily been for cattle feed there is a concern that increased demand for biofuels will encourage soy farmers to expand into the *cerrado*.

In some intensive agricultural systems, such as the EU and the US, land was taken out of agricultural production and set aside to become natural conservation areas. This land is now under consideration for converting back to crop production, although not necessarily for biofuels. There is evidence that farmers in the US are also converting forests into agricultural land (Rathmann et al., 2010). These LUCs could threaten insect and bird populations which play an important role in maintaining the productivity of the agricultural ecosystem.

The introduction of any new species may be problematic if they become invasive, that is they become established and are to expand in natural ecosystems where they have no natural predators and hence they affect biodiversity and ecosystem services. Possible effects could be that an invasive species alters a habitat by being of a different structure (e.g. tree species in grassland), through impacting on fire regimes (for instance by being highly flammable and tolerant of fire), soil fertility (e.g. through fixing nitrogen), and soil hydrology (through sustained high transpiration) (Amezaga et al., 2010). For example, eucalyptus is an exotic species that can have a negative impact on soil water in locations when inappropriate varieties for the soil type are planted (Dessie and Erkossa, 2011). A reason for being concerned about biofuel crops becoming invasive species is that the characteristics that make a good biofuel crop are the same characteristics that make an invasive species: rapid growth, aggressive colonisation of space, ease of establishment, wide habitat tolerance and resistance to pests and diseases. For example, jatropha is considered an invasive weed in South Africa and Australia, although studies on its actual or potential ecological impact under a variety of environmental conditions are as yet unavailable. Indeed, the behaviour of jatropha under a range of conditions appears in general to need further research (Amezaga et al., 2010).

Large monoculture plantations are known to be susceptible to attack by pests which can be the cause of more than 40 per cent of yield losses (Oerke and Dehne, 1997, quoted in Hart Energy Consulting and CIBA, 2010: 17). Oil palms in South East Asia have suffered from basal stem rot, leading to a yield reduction of 25 per cent or more after 10 years (Ariffin et al., 2000 quoted in Hart Energy Consulting and CIBA, 2010: 17). However, there has been little research to quantify biodiversity loss associated with conversion of different habitats to oil palm or the impacts on biodiversity of different management schemes (Donald, 2004).

The impact of biofuels on soil erosion can be positive or negative depending on a number of factors, such as the crop, the soil and cultivation practices (Stromberg *et al.*, 2010). Growing root crops on marginal land implies substantial soil disturbance, which leads to greater risk of soil erosion and release of soil carbon; modern zero tillage methods are preferable but not applicable in the case of root crops. Otherwise, perennial crops with established root systems and low soil disturbance may be more appropriate for marginal conditions where soil will often be poor and loose (Hart Energy Consulting and CIBA, 2010). Perennial crops can help improve soil quality in a variety of ways, such as helping reduce erosion by increasing soil cover, reducing soil disturbance, improving organic matter content and increasing soil biodiversity.

While much attention is given to the ecosystem services that provide food, fuel, fibre and water, less attention is given to cultural services and the role that these play in maintaining social cohesion. If culturally important sites are destroyed by clearing for biofuel then it is possible that this cohesion weakens, and the informal institutions that forbid resource extraction from certain sites will also be weakened. As a consequence both biodiversity and other ecosystem services can be reduced. However, it is not only the demand for biofuels that leads to the exploitation of culturally significant sites. Other pressures, including the commodification of natural resources, which people can use for short-term gain, and the influence of external cultures bringing different value systems, can lead to social tensions particularly between different generations within a community (Shackleton *et al.*, 2008).

Water use and management for biofuel production

This section discusses the effects of biofuel production on water resources from the perspectives of demand and pollution in both the agricultural and the industrial processing phases. The actual effects are site specific and depend on the feedstock used, the agro–climatological conditions of the site where the feedstock is grown, the industrial conversion technologies and water management techniques, as well as on the water availability and consumption patterns in a specific area. Concerns arise from experiences with the commercial crops that can also be used as biofuel feedstocks as well as the withdrawal of large quantities of water affecting the functioning of other ecosystem services.

Brazil, the EU and the US are the largest biofuel producers using a production system of intensive cultivation based on monoculture crops such as sugarcane (for bioethanol in Brazil), maize (for bioethanol in the US) and rapeseed (for biodiesel in the EU). As demand for biofuels grows, this intensive cultivation model is expanding rapidly. Large-scale agricultural plantations use water, fertilisers and agrochemicals to give high yields per hectare. Some biofuel feedstocks have high water demands (see Tables 1.2 and 1.3) which together with chemical run-off can have major ecological consequences particularly for aquatic life. From a social perspective, in regions where water stress exists or is threatened, using water or polluting water resources for biofuel production may exacerbate development problems associated with water scarcity and water resources competition for food crop agriculture, industry and human consumption. It is estimated that about 1.2 billion people live in areas where water is scarce (Comprehensive Assessment of Water Management in Agriculture, 2007).

Water demand

The irrigation water use for biofuel production in 2007 was estimated at 45 billion km^3 or to put in another way, six times more than was used that year globally for drinking water. Biofuels can affect water availability depending on water demand for the agricultural and industrial processing phases of production. This section looks at the two stages separately.

When using product life cycle analysis, the agricultural stage for biofuel production is found to dominate water use (Gerbens-Leenes *et al.*, 2009). Biofuel crops account for $100km^3$ of the water lost to the atmosphere through evapo-transpiration[7] by crops per year at global level (around 1 per cent) and for around 2 per cent of the irrigation water used globally (Comprehensive Assessment of Water Management in Agriculture (CA), 2007). Although this amount appears small compared to the global totals of $7,130km^3$ of water evapo-transpired by other crops and $2,630km^3$/year for irrigation (Molden *et al.*, 2007, quoted in de Fraiture *et al.*, 2008), water used for the production of biofuel crops may have substantial impacts on local water availability depending on the supply – demand ratio and level of rainfall ('green' water) and availability of surface water ('blue' water) in the area of production or downstream if the water enters streams and rivers.

At the agricultural stage of biofuel production, water demand varies with the type of feedstock. For example, sugarcane and maize have high water demands, while sweet sorghum is much less water intensive. It has been estimated that, on average, to produce one litre of bioethanol from sugarcane takes 2,500 litres of water, and one litre of ethanol from maize requires 2,600 litres of water (Gerbens-Leenes *et al.*, 2009). However, there is a significant range depending on the local agro-climatological conditions under which biofuel crops grow. In the US, maize is rain-fed and in China it is irrigated; the water demand per litre of ethanol is estimated at 400 litres and 2,400 litres respectively (de Fraiture *et al.*, 2008). In India, sugarcane requires 3,500 litres of irrigation water to produce 1 litre of ethanol while in Brazil it is substantially less (90 litres) (Hart Energy Consultancy and CIBA, 2010). Generally, biofuel crops grown under rain-fed conditions do not have major impact on water availability unless the water requirements of the crops are higher than the rainfall in the area where they grow. But intensive use of rainfall water may disturb the normal process of blue water generation in the long term (McCornick *et al.*, 2008). This effect, however, requires more research as this issue is hardly mentioned in the literature related to water use for biofuel production.

In contrast to rain-fed feedstock production, irrigated agriculture of biofuel crops may have significant impact on water availability. Irrigation may contribute to water stress downstream of the biofuel plantation and competition with water for other end uses. The actual effects depend on water availability and consumption patterns in a specific location, varying from region to region and from country to country. In Brazil, most of the original sugarcane plantations are rain-fed as the production tends to be limited to regions having a reasonable rainfall spread throughout the year (Moreira, 2007). However, due to the increasing demand for bioethanol, sugarcane is being grown in other regions with less rainfall, where irrigation is required for part of the growing cycle, during drought and, in only a few places, for the whole period (Moreira, 2007). Brazil has plentiful surface and ground-water resources so the water use to water supply ratio is reported to be one of the lowest in the world (Moreira, 2007; Varghese, 2007). Even so, local experiences of water scarcity have been reported in Brazilian bioethanol-producing states. The scarcity has been attributed to poor implementation and enforcement of water use and management policies (Moreira, 2007).

Unlike Brazil, India relies mainly on irrigation for sugarcane production. Approximately 85 per cent of the area planted with sugarcane is irrigated (de Fraiture *et al.*, 2008). India already has water scarcity problems, and cultivation of sugarcane in drought-prone areas has worsened the situation (Tewari *et al.*, 2007). Increasing production of sugarcane to produce bioethanol would exert even more pressure on water resources and may lead to competition for water with food crops (de Fraiture *et al.*, 2008). The effects are fairly localised. For example, three of the top six sugar-producing states in India (Andhra Pradesh, Karnataka and Maharashtra) are part of the Krishna Basin – an inter-river system located in South India. The area of the basin is 259,000km², with 70 million habitants, 68 per cent living in rural areas. It has a per capita renewable water availability lower than the all-India figure (1,130m³ per capita) and the water needs of the local natural ecology of the basin are not often met, especially during drought periods due to increasing water withdrawals and declining water supply (McCornick *et al.*, 2008). The environmental characteristics of the basin thus make sugarcane expansion for bioethanol production unsustainable. Taking into account the potential negative effects on water resources of expanding sugarcane for bioethanol production, India currently allows bioethanol production only from molasses – a by-product of the sugar processing (Zhou and Thomson, 2009).

It has been estimated that the increased global demand for irrigation water to meet the planned expansion of biofuels (as of 2007) would be 180km³ (de Fraiture *et al.*, 2008) which is regarded as only a small additional increase compared to that for food crops. However, it would appear that it is local impacts that are likely to be the most significant, potentially in India and

China, and even in these countries there will be considerable regional variations. If there is a switch to secondary or tertiary biofuel crops then the water demand picture may well change. Indeed, if ligno-cellulose from tree crops becomes significant, then it is possible that water availability may improve due to the positive effect that perennial tree crops can have on soil water retention and reducing run-off (Berndes, 2002). However, if these are exotic species, then it is possible that they can have a higher water demand than the local natural vegetation (Amezaga *et al.*, 2010).

Water requirements for industrial processing of biofuels are considerably less than for feedstock production (Moreira, 2007; Varghese, 2007; Schnoor *et al.*, 2008). However, since the industrial process is confined to a small area, the local effects can be significant, depending on the quantity of biofuel produced, the nature of water management techniques used and the extent of water recycling adopted by the specific mill/biorefinery, water availability and demand in the catchment area. For example, bioethanol producers in the US and Brazil have increasingly implemented water use and water recycling measures during the industrial process (Moreira, 2007; Varghese, 2007; Schnoor *et al.*, 2008). In the US, where bioethanol is mainly produced from maize, 7 gallons (26.5 litres) of water are used per gallon (3.8 litres) of ethanol and with improved water management the figures are reported at about 4 gallons (15.1 litres) per gallon of ethanol (Schnoor *et al.*, 2008). In São Paulo State, where almost 60 per cent of the ethanol production in Brazil is concentrated (Walter *et al.*, 2008), water collection per tonne of cane for processing sugarcane decreased from $5.6m^3$ in 1990 to $1.83m^3$ in 2005. However, in the context of Brazil, it is difficult to establish the precise water use for ethanol production at the industrial phase because most of the mills maintain the option to switch output between sugar and bioethanol, and there are stages of the industrial process commonly applied to both products. In contrast, in India water requirements in molasses distilleries are still a major concern, bearing in mind water scarcities in the country. Indian molasses distilleries use approximately 36.5×10^3 litre of water per 10^3 litre of bioethanol produced but there are initiatives to reduce water use during the industrial stage (Tewari *et al.*, 2007).

The stress on water resources produced by biorefining also depends on the availability of water in the area and the scale of production. For example, a small-scale processing plant can have a low level of withdrawals but if it is located in an area with limited water resources withdrawal can still exacerbate water scarcity. By contrast, in areas with good levels of water availability, a processing plant producing large amount of biofuels and/or using large quantities for fresh water in the industrial process may not lead to local water shortages.

Water pollution risks

During the agricultural stage the major risk of water pollution comes from fertilisers and agrochemicals (pesticides, herbicides and fungicides) run-off which contaminates open water systems and underground water. In addition, monoculture plantations are prone to soil erosion and nutrient run-off to surface water. Extensive use of fertilisers and pesticides, required to maintain high levels of production, leads to increase of chemical nutrients in aquatic ecosystems causing eutrophication, lack of oxygen and hence severe reductions in water quality, fish and other animal populations (Zhou and Thomson, 2009). For example, in the US, the extensive application of fertilisers to maize grown in the Mississippi basin is a major contributor to nutrients found in the Mississippi River (National Academy of Sciences, 2008; Dominguez-Faus *et al.*, 2009). In 2008 a consequence of nutrient discharge from the Mississippi River into the Gulf of Mexico was that an area of more than 20,700km^2 suffered from eutrophication (Dominguez-Faus *et al.*, 2009). Therefore, there is a major concern that increases in maize-based bioethanol production may lead to increased eutrophication in and around the coastal waters of the US (National Academy of Sciences, 2008).

Conversely, in São Paulo State, Brazil, the production of sugarcane is considered to have little impact on water quality according to an evaluation carried out by the Brazilian Agricultural Research Corporation (EMPRAPA) (Moreira, 2007). Overall, the quantity of agrochemicals used for sugarcane and the rates of fertiliser application are less than those used for other crops in the country. In addition, the risk of water contamination due to chemical nutrients is lower because mineral fertilisers are supplemented by vinasse – the liquid waste from sugar and bioethanol production which is acidic and rich in nutrients (Moreira, 2007). However, vinasse is a highly polluting effluent (see next paragraph). Therefore, although this alternative use of vinasse represents less environmental risk than the use of agrochemical fertilisers, the impacts on soil chemistry of its long-term application require more research.

At the industrial processing phase the effluents generated have a great polluting potential. How these effluents are managed significantly affects quality of water resources. Disposal of vinasse in open water systems can cause major water contamination with devastating effects on aquatic ecosystems and make the water unsuitable for other uses (Moreira, 2007). The impacts differ between countries. Brazil has a 30-year history of dealing with management of effluents from ethanol distilleries and achieving improvements in vinasse treatment and water recycling; whereas in India contamination of water caused by molasses-based alcohol distilleries is still a major problem. In the first years of the Proálcool programme[8] in Brazil there was considerable environmental damage due to the uncontrolled dumping of vinasse in watercourses, but the adoption of regulations and the use of vinasse for both fertilisation and irrigation are considered to be significant factors in the management of the pollution reduction (Moreira, 2007). In India there are regulations in place for distilleries to have achieved zero discharge in inland surface watercourses by the end of 2005 (Tewari *et al.*, 2007: 352). Distilleries are still working on compliance by applying some of the wastewater treatment methods recommended by the governmental agency dealing with environmental issues. It has been reported that, as of 2007, over 50 per cent of the distilleries in the major bioethanol-producing states of Maharashtra and Uttar Pradesh had complied with the regulation (Tewari *et al.*, 2007).

The polluting effects of biofuel production in the agricultural stage vary according to the crop but in general they can be lessened by using agricultural techniques of ferti-irrigation, no-tillage and conservation of the soil to avoid erosion (Moreria, 2007). Equally, contamination caused by effluents can be minimised and almost reduced to zero by using enhanced recycling and wastewater treatment techniques. To achieve these goals, the adoption and strict enforcement of the proper regulation is required as well as the producer's commitment to implementation of the necessary changes. The case of Brazil represents both good and bad practices. In the first years of the Proálcool programme much damage was caused to water resources by allowing unrestricted dumping of effluent into waterways. However, the development of vinasse for irrigation and improvements in water management has led to a situation where water contamination from ethanol production is no longer considered the norm.

Palm oil mill effluent is characterised by high levels of organic matter which, if the waste is released into waterways, results in the depletion of dissolved oxygen levels to such an extent it can lead to eutrophication. In Malaysia, it is estimated that the quantity of mill waste produced is equivalent to the domestic sewage produced by 1.5 million people (Donald, 2004). Technology for waste treatment exists, although it appears that leaks are particularly prone from small mills (Sheil *et al.*, 2009).

Ecosystems-friendly biofuels

In respect of biofuels expansion, the concern related to ecosystems and their associated services, as to what might happen if habitats are destroyed by large-scale planting of biofuels is based on experiences from the development of other sectors, such as large-scale commercial agriculture.

The threat from biofuels to ecosystem habitats and the associated biodiversity arises from land-use changes under three scenarios:

1   when land not currently used for commercial crops, e.g. woodland or savannah, is brought into production
2   'degraded' or waste land is brought into production
3   when one crop is substituted for another.

(In this chapter we have focused on the impacts on biodiversity rather than on food availability which is the subject of the next chapter.) Indirect land-use changes can also occur when biofuels displace rural people onto more marginal land where cultivation or grazing will affect biodiversity and ecosystems (Cotula *et al.*, 2008).

Biofuels is only one of multiple drivers of ecosystem change. There are natural drivers such as climate change, volcanos and tsunamis. There are socio-economic drivers such as commercial logging, forest systems change, land conversion for agriculture, building dams and expanding cities. These drivers are brought about by factors such as prices, technology change and public investment which are outside the control of the rural poor. In this section we first briefly review factors of logging, climate change and petroleum extraction for their impacts on ecosystems, as well describing how attempts at biodiversity conservation have been detrimental to poor people's access to ecosystem services. We then examine how biofuels can be produced in an ecosystems-friendly way, both from agricultural and institutional perspectives.

Non-biological drivers of ecological systems change

Logging

Political inertia, competing priorities and lack of capacity and understanding, not to mention high levels of demand for timber, often make it attractive to clear forests and to build roads. Indeed, it is estimated that, as of 2000, less than 1Mha out of 5.3Mha of land allocated to palm oil development in Kalimantan (Indonesia) had been planted with the palm (Casson, 2000). Here, licences to establish palm oil estates had provided a loophole for companies who had no intention of planting trees but wanted to extract the tropical timber for short-term profit, knowing full well that they would not get a licence for timber felling. Illegal logging is reported to be a prime cause of the depletion of natural common property resources which leads to the destruction of biodiversity and ecosystem services (WRI *et al.*, 2005).

Climate change

Greenhouse gas-induced climate change is predicted to bring major changes in ecosystems accompanied by potentially significant loss of biodiversity (Fischlin *et al.*, 2007). By the end of the 21st century, climate change is likely to be the major driver of biodiversity loss (Swiderska *et al.*, 2008). Changes in temperature and rainfall patterns are likely to displace habitats and some habitats may disappear altogether leading to species extinction (Thomas *et al.*, 2004). Because species distribution is often a function of climate, e.g. tropical rainforest only occurs under warm humid conditions, then alterations in basic parameters such as temperature and rainfall patterns will in turn lead to changes in the occurrence of plants and animals and hence overall levels of biodiversity. Models of the effect of climate change on biodiversity controversially predict high rates of extinction. For example, a study carried out in 2004 using a sample of 1,103 terrestrial plants and animals suggests that 15–37 per cent would lose their climatically determined habitat by 2050 (Thomas *et al.*, 2004). Moreover, climate change is associated with major regional variations in the hydrological cycle with an increase of extreme events such as floods and droughts, which also impact on habitats.

Petroleum extraction

Considerable damage to habitats in terms of loss or fragmentation occurs in the normal course of oil extraction. The release of toxic chemicals, e.g. arsenic and benzene, has negative impacts on animals and plants (Worldwatch Institute, 2007). The extraction techniques used can also damage aquifers. There can be land enclosure to protect wells, which restricts access to the local community, while pipelines can disrupt routes of people and animals. Offshore drilling can lead to the destruction of marine ecosystems through dredging. Oil spills from land-based wells have been reduced compared to the early days of prospecting; but it is more difficult to contain oil leaks into marine environments, which can cover an extensive area. For example, the slick caused by the oil leak from the *Deepwater Horizon* well in the Gulf of Mexico in 2010 covered an area of 180,000km^2 and affected up to 790km of coastline in the southern United States.[9] The oil and some of the methods used to try to disperse the oil have negative impacts on ecosystems and cause damage to livelihoods such as fishing and tourism.

There has been considerable concern expressed about the potential ecological damage that can occur from the exploitation of tar sands. Reports state that extraction from tar sands in Canada has already led to a loss of a diversity of habitats such as forests, shrub lands, water bodies, bog and fen wetlands, as well as affecting water quality (Timoney and Lee, 2009). Only a small number of countries (Venezuela, the Congo and Madagascar) in the South are considering exploiting their tar sand reserves.

Biodiversity conservation

It must also be remembered that agricultural or other forms of landscape transformation, with their associated impacts on biodiversity, are not the only land-use change that can seriously affect poor people's access to ecosystem services. Indeed measures taken to protect biodiversity can also have negative impacts on livelihoods of the rural poor, particularly where conservation efforts are geared towards large carnivores and herbivores, e.g. lions, tigers, elephants and rhinos, which can involve alienation of large tracts of land from local communities. For example, the Selous Game Reserve, in Tanzania, covers 55,000km^2, and no permanent human habitation is allowed, despite the land being suitable for agriculture. Game reserves in East Africa are the subject of long-running human rights concerns of indigenous groups (Mwaikusa, 1993; Neumann, 1995; Wanitzek and Sippel, 1998). These conservation measures may preserve biodiversity and help maintain ecosystem services but in so doing they create a divide between humans and nature, dispossessing traditional users of the natural resources, reducing their use of ecosystem services and increasing their vulnerability (Swiderska et al., 2008).

Conservationists have attempted to alleviate social impacts by involving local communities with integrated conservation and development projects so that people living around parks and reserves can be part of the economic benefits derived from, for example, tourism or hunting (WRI *et al.*, 2005; Swiderska *et al.*, 2008). There is now sufficient experience with community-based natural resource management that the broad principles for success are known but the major barrier to successfully balancing biodiversity conservation and livelihoods is related to governance issues (Shackleton *et al.*, 2008).

Ecologically sustainable biofuel production

Although our knowledge is limited of how biodiversity and ecosystems function, there are some general guidelines that could be used to protect them when deciding whether or not an area should be used for biofuel crops. Three parameters that could be used to assess whether or not biofuel production could be carried out in an ecologically sustainable way, protecting biodiversity and ecosystems services are: (a) where it is planted, (b) how it is grown, and (c) the type of biofuel crop grown.

Biodiversity, as measured by numbers of species, tends to be well defined geographically or located in particular types of landscapes, which can be translated into avoiding habitats such as natural and semi-natural forests, grasslands, wetlands, mangrove swamps and peat lands and particularly those areas labelled as biodiversity 'hotspots'.[10] It is also a fairly safe assumption that anywhere with equitable temperatures, a relatively high rainfall and high degree of variation in species, and which still retains a good cover of native vegetation, will also be rich in biodiversity. At the simplest level, these areas should be given priority for preservation from development for biofuel plantations. Other land area types to avoid for biofuel production include those where there is water stress and in watershed areas. This avoids damaging ecosystem services based on water vital for agriculture as well as human and animal well-being. Land classified as 'marginal' is not only likely to suffer from water shortages but is also likely to be susceptible to erosion and hence bring further diminution of the ecosystem (Cotula *et al.*, 2008).

The management practices employed in growing biofuels are important for sustaining biodiversity and ecosystems services. Solutions for ecologically sustainable first generation biofuel crops are the same as those for food crop agriculture: low inputs of fertilisers derived from fossil fuels and avoidance of monocultures. Knowledge about how to carry out low input agriculture without damaging levels and quality of outputs is increasing (Swinton *et al.*, 2007). Intercropping of grain and legumes has long been known to give good yields without the need for high levels of chemical fertilisers (Snapp *et al.*, 2010).

Planting patterns that retain areas of linked natural vegetation ('wildlife corridors') are considered to help maintain biodiversity. In Brazil, the state of São Paulo requires sugarcane growers to set aside 20 per cent of their total planted area as natural reserves (UN-Energy, 2007). Such approaches in oil palm plantations have shown that the natural vegetation can attract insect predators of species which attack oil palms hence reducing the need for chemical insecticides (Stromberg *et al.*, 2010). The Indian Hassan Biofuels project is encouraging small-scale farmers to plant mixes of native species of non-edible oilseed-bearing shrubs and trees for biodiesel production in hedges that line field bunds and act as shade trees. This is a deliberate strategy of directly combining biodiversity and biodiesel production and builds upon traditional practice (Narayanaswamy, 2009). Where this type of agriculture is carried out on degraded land, there is the potential to upgrade the ecosystem, increase biodiversity and improve ecosystem services, for example, soil quality (UN-Energy, 2007).

The type of crop grown can have considerable impact on biodiversity. Perennial crops are more likely to have an ecosystem that is more biodiverse than that of an annual crop. The conversion of pasture lands to annual crops has considerable negative ecological impacts. The type of crop also influences the yield per hectare: the higher the yield the reduced demand for land for a given output and hence reduced threat to biodiversity. There is a considerable variation in yield per hectare between crops. For ethanol crops the sugar yield range is between 1.3 tonnes per hectare (sorghum) and 100 tonnes per hectare (sugar beet) (see Table 1.2) while for biodiesel the range is between 199kg per hectare (soy) and 5,000kg per hectare (palm oil) of oil (see Table 1.3).

Another approach is to reduce the demand for land needed for growing the crops by increasing sugar and oil yield efficiency. Improving efficiency can be both in terms of maximising the amount of feedstock harvested per hectare and the amount of sugar or oil extracted per unit weight of feedstock. A number of plants being considered as biofuel feedstocks are not currently grown commercially under a range of environmental conditions, therefore extensive plant breeding to produce higher and consistent yields under a variety of conditions are needed (however, as pointed out above, this might have negative consequences for biodiversity particularly when switching to 'second generation' crops which could become invasive).

The potential of low water demanding feedstock

In this chapter we have highlighted the water use of first generation biofuel crops because of the potentially serious impacts on water ecosystem services. It is also one of the areas where there is a body of empirical evidence primarily due to the disposal problems of waste from the processing phase of both sugarcane and palm oil. As shown above, many of the biofuel crops also have a high water demand, which has been driving interest in a search for potential feedstock crops with a low water demand. Two factors have generated an interest in two specific crops: jatropha (biodiesel) and sweet sorghum (ethanol). First, there is the aim to incorporate small-scale farmers into the biofuel value chain as these biofuel crops are considered appropriate for small-scale production. If the so-called 'waste land', which is too dry for commercial agriculture and does not have clear tenure rights, is also brought into production then landless people could also potentially participate. Second, their low water demand does not necessarily put them into direct competition for water for food production, so they become an attractive option in semi-arid regions or areas where there is regular risk of drought.

It is estimated that jatropha and sweet sorghum have higher yields per unit of water than other biofuel feedstocks such as sugarcane, maize or soy (ICRISAT, 2007; Rajagopal, 2008). Jatropha is a widespread tropical shrub producing non-edible oilseeds that can be used for biodiesel production. In its natural habitat it has low water requirements and in cultivation adapts well to marginal soils with low nutrient content. In addition it is estimated to produce a higher biofuel yield per unit of water than the existing annual oilseed crops (ICRISAT, 2007; Rajagopal, 2008). However, evidence available on biodiesel production from jatropha is limited to small-scale production trials or at village level when grown for local consumption. A jatropha biodiesel project carried out in Mali between 1987 and 1997 showed that yields are not predictable (Fairless, 2007; Henning, 2009) and there is a large variation in estimates of potential production (ICRISAT, 2007).

In consequence, although there has been great optimism about jatropha's potential for producing a biofuel that does not compete with food crops, there is still uncertainty about the oil yield under conditions of poor land quality and water scarcity; and this has raised questions about its economic viability. There is a risk that to obtain high economic yields more water and better land would be necessary, thereby defeating the object of using jatropha in the first place (Fairless, 2007; ICRISAT, 2007; Jongschaap *et al.*, 2007; Rajagopal, 2008). An example of the consequences of poor agronomic data for jatropha has been already reported in Swaziland, where farmers were involved in a project to grow jatropha for biodiesel production under outgrower contracts to a foreign company. Some farmers complained that contrary to what they were told 'Jatropha needs to be watered once or even three times a week and that water collection for jatropha crop is competing with collection for domestic use such as cooking and sanitation' (Burley and Griffiths, 2009).

Sweet sorghum is also a promising feedstock for ethanol production. Sweet sorghum is a relative of grain sorghum with the difference that it stores sugar in the stalks and at the same time giving reasonable grain yields (ICRISAT, 2007). It is considered to have wide adaptability, resistance to drought and saline-alkaline soils and tolerance to waterlogging (Rajagopal, 2008). Because of its suitability for dry and semi-arid lands and its estimated high net energy balance under tropical conditions, the International Crops Research Institute for the Semi-Arid Tropics (ICRISAT) is promoting sweet sorghum for ethanol production in Asia and Africa (Reddy *et al.*, 2008). The main evidence available at the time of writing related to bioethanol from sweet sorghum comes from India. In Andhra Pradesh, a project based on 3,200 small-scale farmers holding 2ha size (approximately) has shown that sweet sorghum uses four times less water than sugarcane. In addition, the results of studies in India demonstrate that sweet sorghum has a higher ethanol yield per unit of water (mm) than sugarcane: 3.45 vs. 1.65 (Rajagopal, 2008). However, because the sweet sorghum is not yet under large-scale commercial production, there is still considerable uncertainty with regard to the water requirements of the crop when it is grown under large-scale plantation conditions.

Finally, another potential advantage attributed to bioethanol from sweet sorghum is that the wastewater effluent from the industrial process seems to be less polluting than that released during the production of ethanol from sugarcane molasses (ICRISAT, 2007).

Institutions and governance

Land-use changes influence ecosystem services and human well-being. LUCs are mediated by institutions, both formal and informal. Decision makers at different levels directly or indirectly affect ecosystems and their services. Certainly the decisions they make about biofuels have impacts. When property rights to local ecosystems are ill-defined or inadequately protected, then ecosystem services are at risk and the poor are at risk of displacement to more fragile ecosystems. Who has access and control will determine who benefits from the services. When institutions are weak and ineffective, it is likely that the poor will lose out. Power relations ensure that certain groups are able to capture and control formal institutions, or at least ensure that they do not function effectively. This can be seen in biofuels where powerful outsiders are able to enter alliances with government officials to gain access to land and exclude the poor (Cotula *et al.*, 2008). Local needs are frequently overridden by outsiders' demands. This effect is not confined to biofuels. In terms of biodiversity conservation, it is global values that dominate which are primarily non-use values, such as the protection of iconic species (e.g. tigers) as opposed to the local utilitarian values (food, fuel, medicines) (WRI *et al.*, 2005).

Another example of unequal power relations comes from the approach biofuel developers have adopted to avoid criticism of clearing forests. They have turned to bringing other types of land into biofuel production, most notably land designated as 'degraded'. The use of 'degraded' land involves a decision about what constitutes degraded and also importantly who decides what is degraded. For example, a forest that has been subject to logging can still support considerable levels of biodiversity, albeit lower than before logging began, and can continue to be a source of valuable ecosystem services. The designation of 'degraded' is generally done by government officials without consultation with local communities (Hart Energy Consulting and CABI, 2010). As was pointed out in Chapter 3, this is not necessarily how rural people, particularly the poor and more vulnerable, will view this land – instead they see it as a valuable source of goods and services – a situation which can be adequately summarised as: 'degraded how is related to degraded for whom, and regenerate how is related to regenerate for whom' (Ariza-Montobbio *et al.*, 2010).

Little attention is given to the importance of cultural services in rural planning, and informal institutions that govern cultural attitudes to natural systems are generally ignored (Shackleton *et al.*, 2008). However, local knowledge about ecosystems, their management and cultural attitudes and beliefs can make an important contribution to maintaining biodiversity and the services it supports. Such an approach has been incorporated into the national biodiversity strategy in Mozambique (Virtanen, 2002, cited in Shackleton *et al.*, 2008: 69).

Market mechanisms do not always protect ecosystems services because many services do not have market values. The establishment of large privately owned biofuels plantations has the potential to accelerate the process of land privatisation, leading to the loss of communal control over natural resources that have until now been part of common property. The threat to ecosystem services from biofuels has very similar causes to those described in Chapter 3 in relation to land used for agriculture: power asymmetries and a lack of recognition of traditional law and practice. There is evidence to show that rural communities also develop rules for the management of natural resources, such as forests, based on accumulated knowledge of the way their ecosystems work (Laumonier *et al.*, 2008). These management systems help preserve biodiversity and ecosystem services (Swinton *et al.*, 2007). Unfortunately, this traditional knowledge is ignored or rejected by outsiders.

Preserving biodiversity is both an ethical consideration (all living things have an intrinsic value) and an economic consideration although the perspectives are rooted in different discourses and hence can be difficult to reconcile. Trying to bridge the gap is part of the governance of the biofuel value chain which we will discuss in Chapter 7. There is some pessimism about whether or not it is possible to address poverty and preserve biodiversity simultaneously (Shackleton *et al.*, 2008). The situation is made complex by the fact that value given to biodiversity can be different from cultural, agricultural and ecological perspectives, while ecosystem services will be valued differently by different groups, at different times, in different locations reflecting societal choices and values. It should also be kept in mind not to focus exclusively on maintaining biodiversity since not all ecosystem services are derived from biodiversity, e.g. water flow (Haines-Young and Potschin, 2010). For those ecosystems threatened by forces of globalisation, such as biofuels to meet transportation demands, the solutions are complex because of the numbers of stakeholders and the different levels at which they operate, increasing the complexity of the types of interventions needed.

At the company level, it is possible that awareness raising and appealing to corporate social responsibility might be effective (see Chapter 6). Stakeholders who are not involved in environmental issues, or are outsiders unaware of the composition or relevance of a specific ecosystem to communities, are probably not aware of the effects that their activities have on biodiversity and ecosystem services, and the consequences of those effects – particularly on the poor. Large-scale growers of biofuel crops need to be persuaded of the necessity to adopt farming management systems that protect ecosystem services. This needs good motivation since the services' beneficiaries might be located elsewhere to the place of production. In Australia, schemes have been tested which provide financial credits for biodiversity preservation (Landell-Mills and Porras, 2002), although it was not clear how such a scheme could ensure that the poor benefit and what sort of measures would be used to enforce agreements. Although such financial credits might be labelled 'subsidies' and resisted in some quarters, such outlays can be seen as paying for a service. In the US, it was found that providing farmers with incentives to carry out land/water management offers a cheaper method of supplying drinking water than using treatment plants (Bennett *et al.*, 2005).

The inclusion of an assessment of the impacts on biodiversity and ecosystem services when a large-scale biofuel project is being proposed would seem logical and could be included under sustainability criteria. However, there is limited capacity and understanding about how ecosystems function and the importance of ecosystem services particularly for the poor, which makes informed policy formulation difficult. The process is not helped by the need for integrated, cross-sector coordination. For example, agricultural policies, which are likely to govern first generation biofuel crop production, in the South are often not aligned with policies and programmes to promote biodiversity. There is also a lack of understanding of how to integrate knowledge about ecosystem functioning into natural resource management, limiting the potential to minimise any trade-offs between preserving habitats and development (Laumonier *et al.*, 2008). There are only now beginning to emerge tools and methodologies for carrying out such an assessment (FAO, 2010; Global Bioenergy Parntership, 2011).

Can biofuels be produced in an ecosystems-friendly way and also be pro-poor?

The dependence of the poor on natural systems for the provision of basic goods and services, as well as a source of income, is increasingly recognised, hence why the Millennium Ecosystem Assessment was established by the UN. Ecosystem services are the safety net in times of crisis for many people in the South who have no access to state or private insurance schemes. However, we have limited understanding of ecosystems, their functioning to provide services and the impact on poor people's livelihoods. Nevertheless, when ecosystems are threatened there is a concern that vulnerability of the poor increases if the level and quality of ecosystem services are reduced and/ or they are denied access to areas from where they have been able to draw on services (World Bank, 2008a).

At the beginning of this chapter we posed the question as to whether or not biofuels could be produced in an ecosystem-friendly way that was also not detrimental to the poor in terms of reducing their access to ecosystem services. Pro-poor biofuels would ensure that the ecosystem services used by the poor are not damaged by the agricultural system to grow biofuel crops and/or the processing system to produce the fuel. Although we have limited scientific knowledge about the functioning of ecosystems as well as the role of these systems in the lives of the poor, we do have sufficient knowledge to make sensible suggestions about where not to grow biofuels so that biodiversity and the associated ecosystem services are not damaged. There also exist methods for growing and processing crops which can be considered ecosystems friendly. However, there does need to be more work done on improving the yields of a range of plants under a variety of soil and climatic conditions so that informed decisions can be made. Nevertheless, unless the institutions that mediate agricultural development take a more pro-poor stance, the rural poor will remain vulnerable to large-sale commercial biofuel production.

In this chapter we have looked at the impact of large-scale biofuels expansion on ecosystems and the consequences for the derived services used by the rural poor. Without the proper safeguards these can be quite significant, although these effects are probably less generally recognised than the issue we will address in the next chapter: biofuels and the food security of the poor.

5   Liquid biofuel production and rural communities' food security

Introduction

A great portion of the world's poor live in rural areas and rely on agriculture and land-based natural resources as a means of living (see Chapter 2). According to the UN Millennium Project Task Force on Hunger, half of the world's hungry live in smallholder farming households, around one-tenth are pastoralists, fishermen and forest users, and approximately one-fifth are rural landless (UN Millennium Project, 2005: 3–4). Therefore, the impact of biofuels on the food security of the rural poor is of utmost interest: do biofuels increase rural incomes and enable more people to buy food? Or do biofuels displace staple crops as well as resulting in the loss of communal land that forms an important resource for the poor?

The general press, non-governmental organisations, officials of international organisations and members of the scientific community have voiced strong concerns about the effects that the increased demand for biofuels may have on food security through the competition for agricultural land between biofuel and food crops. Price increases of food staples in 2007 and 2008 have become inextricably linked to the interest in maize and oilseeds for biofuel production. In Chapter 1 we referred to the concerns of the former UN Special Rapporteur on the Right to Food about using good quality agricultural land for growing biofuels. Describing such an action as 'a crime against humanity' is very emotive language, but are the claims justified? This chapter assesses the evidence as to whether or not biofuels are really displacing staple crops and driving up staple food prices.

Food security is a complex concept incorporating more than availability and price. Discussions based on a simplistic understanding of what food security entails run the risk of a superficial analysis and drawing misleading conclusions. This chapter begins, therefore, by describing how food security is understood by the development community and why it is an issue of central importance for many development agencies. Then, the extent to which biofuel production influences food security is discussed. It also examines the connections between poor people's food security and the effects of plantation biofuel production on access to land and land-use changes linked to food availability. The chapter concludes with an analysis of whether or not biofuel production can be pro-food security.

The concept of food security

Hunger is considered to be the leading threat to global health, being regarded as responsible for more deaths than HIV/AIDS, malaria and tuberculosis combined (UN-Energy, 2007). It is estimated that up to 25,000 people, two-thirds of them under-five-year-old children, die *each day* from hunger. This statistic underlines, in addition to the moral position, the reason why addressing hunger receives so much attention from the international community in preventing avoidable loss of life.

The UN agencies and many other development organisations consider that hunger can be best addressed through creating conditions of food security. This concept was defined in the 1996 World Food Summit Plan of Action which states that food security exists when all people, at all times, have physical and economic access to sufficient, safe and nutritious food for a healthy and active life (FAO, 1996).[1] This statement reflects Article 25 of the Universal Declaration of Human Rights:

> Everyone has the right to a standard of living adequate for the health and well-being of himself and of his family, including food, clothing, housing and medical care and necessary social services, and the right to security in the event of unemployment, sickness, disability, widowhood, old age or other lack of livelihood in circumstances beyond his control.[2]

The Plan of Action is used by the international development agencies, in particular the FAO, to provide a framework for their work to create food security and to tackle hunger. The Plan of Action conceives of food security as having four dimensions: food availability; stability of supply and access; access to food; and food utilisation (see Box 5.1 for definitions). Interestingly, there is no mention of energy which is needed to cook food and to boil water (which can be a necessary action to make the water safe to drink). The Plan is linked to the Millennium Development Goals of which Goal 1 aims to halve world hunger by 2015.

The international development organisations are target orientated. They use quantitative indicators to measure progress in meeting these targets. This makes a concept like 'hunger' very difficult for them to work with since it is a subjective notion and difficult to quantify. It is also a highly emotive term. Instead the international agencies prefer to use the concept of food security which involves four conditions that are quantifiable:

- adequacy of food supply or availability
- stability of supply, without fluctuations or shortages from season to season or from year to year
- accessibility to food or affordability
- quality and safety of food.

The quality of food available is particularly important since it is the nutritional value of food that is particularly important in determining human

---

Box 5.1 Dimensions of food security as defined by The World Food Summit Plan of Action

Food availability means that the supply of sufficient quantities of food of appropriate qualities, through domestic production or imports (including food aid) is guaranteed.

Stability of supply and access is connected to price fluctuations, weather conditions, human induced disasters as well as a variety of political and economic factors.

Access to food is determined by the level of people's assets (such as land or income) that allows them to grow or buy a variety of appropriate foods for a nutritious diet.

Utilisation of food refers to the ability to make proper use of food through an adequate diet, clean water, sanitation and health care.

Source: Broca, 2002: 6; FAO, 2008b: 72.

---

welfare. Trace elements, vitamins and minerals are essential for people to live a healthy life. For example, about one billion people are estimated to have an iron-deficient diet. Iron deficiency anaemia is considered to be responsible for 20 per cent of global maternal mortality (Cohen *et al.*, 2008: 1). An intake of food with insufficient calories and other essential nutrients impairs mental and physical development in children, while in adults it affects the capacity to work both in terms of the number of days lost due to sickness and in terms of reducing stamina to sustain the rate and the period of working. However, food is not the only input required for good health: clean water and proper sanitation are also necessary components. The way in which people access food is also an important part of their well-being. In Chapter 2, it was mentioned that poor people feel that they lack respect which could occur if they have to rely on food distributed by others or through socially unacceptable coping strategies, such as scavenging or stealing (Anderson, 1990).

Food insecurity can be considered the converse of food security and is generally understood as the limited access to sufficient quantities of nutritionally adequate and safe food. It is estimated that around half of all the people suffering from food insecurity are small-scale farmers (Cohen *et al.*, 2008). While hunger, in the sense of 'recurrent and involuntary lack of access to food' (Anderson, 1990: 1598) is associated with food insecurity, hunger is not always an outcome of food insecurity. This recurrent lack of food leads to the physiological condition of undernourishment when a person's calorific intake is below the minimum dietary energy requirement for light work (FAO, 2009: 8). In this sense hunger, when equated with undernourishment, is detrimental to health. Households and individuals can experience food insecurity in different ways. For some it is a long-term condition when the household or individual is suffering from *chronic food insecurity*;[3] while for others it is a short-term condition resulting from an unexpected or short-term reoccurring event, e.g. illness, crop failure, seasonal scarcities, unemployment, when the household or individual is suffering from *transitory food insecurity.* A household or individual is considered food secure only when they are not vulnerable to both types of insecurity (Rahman Osmani, 2010). It is not only ending chronic food insecurity that needs to be addressed but also the situation where households or individuals in transitory food insecurity slip into chronic food insecurity.

Food insecurity and ill health are generally linked to poverty. Responses to addressing these interlinked conditions have been influenced by the work of the Nobel Prize winning economist, Amartya Sen,[4] who considered that at the individual or household level food security is mediated by the 'substantive freedom of the individual and the family to establish ownership over an adequate amount of food, and this can be done either by growing the food oneself (as peasants do), or by buying it in the market (as the non-growers do)' (Sen, 2001: 161). In other words, if people have sufficient land or income they can produce or buy the nutritional food they need and their food insecurity diminishes. Sen identified three factors that influence access to food: ownership of production resources, available technology of production and level of income obtained through the sale of goods or participating in the labour force (Sen, 2001, quoted in Clancy, 2008: 419). Addressing these factors can ensure that households and individuals are able to avoid chronic food insecurity and are not vulnerable to transitory food insecurity. The quantity and quality of these assets must be sufficient to ensure food security. A household must own enough land, with enough inputs (both labour and agricultural) or receive wages of a level to enable access to food. Sen also points to another important factor in ensuring food security: who in the household controls the resources? There is a large body of evidence to show that 'the greater the degree of control exercised by women over the family income, the greater the proportion of income spent on food' (Rahman Osmani, 2010).

Once a household has acquired food, household members need to benefit from the nutritional value. Here storage (which is important also in safeguarding against transitory food insecurity), preparation, in particular under hygienic conditions, and cooking play a role in ensuring palatable food. Storage in rural households is generally inadequate and there can be substantial loses in quality and quantity of food over time. Under the gender division of labour in households, food preparation is generally the task of women (although there can be cultural differences). Women's time pressures can be a constraint to do the task adequately (see Chapter 2).

Therefore, to assess the impact that growing biofuel crops can have on food security/insecurity, we need to look at the evidence about the influence that the global demand for these crops has had on food availability, including stability of supply, and access to food, not only in terms of quantity but also the nutritional quality, for poor households. If the arguments that biofuels bring rural development are correct, then rural household incomes should increase through growing biofuels, owning processing plants or being employed in the supply chain. An important issue in terms of food security relates to Sen's point about who is earning the income: are women benefiting financially from participating in biofuel value chains?

Food availability

At the global aggregate level, availability of food has not been a major concern in the international community since agricultural production has been rising from 1990 to 2006 (being the time period for which comprehensive data is available at the time of writing) and has managed to keep pace with population growth. Indeed FAO considered that in 2010 there was enough food in the world to feed everyone. At the current level of world food production it has been estimated that there is enough to provide the existing world population with 2,700 calories a day (Evans, 2009: 34). This figure exceeds FAO's cut-off point for determining the hunger threshold which ranges from 1,600–2,000 kilocalories per person per day, depending on the age and gender distribution in each country (FAO, 2008a: 8). However, at the individual level, access to food is a major problem. According to the FAO the number of chronically hungry people increased by 75 million in 2007 reaching a total of 923 million (FAO, 2008a: 4), during a period in which 'the world has grown richer and produced more food than ever in the last decade' (FAO, 2008a: 4). Sixty per cent of the chronically hungry live in either sub-Saharan Africa or South Asia (Cohen *et al.*, 2008: 14). A substantial proportion of these people rely on food aid from international agencies such as the World Food Programme or Oxfam. These organisations buy the food on the open market so their capacity to help the most needed can be strained when food prices rise (see next section below).

Despite the global good news, two issues have begun to cause concern in international development agencies because of the potential effects on food insecurity: a decline in food output and an increase in food prices. A number of factors have contributed to these phenomena. In the least developed countries output declined in 2006 after nearly a decade of modest growth (FAO, 2008a). It is in these countries where a large portion of the world's hungry live.[5] There was also a 20 per cent production shortfall in two of the world's major cereal exporters, Australia and Canada, which were caused by droughts linked to climate change (FAO, 2008a: 42). These two countries, together with Brazil and Argentina, account for between 35 and 40 per cent of world exports so any disruptions in their exports have significant implications for food availability and food prices on the international market (FAO, 2008a: 105). There has also been a decline in public investment in agricultural research which is considered important for the support to smallholder farmers who are not of particular interest to the commercial sector despite being responsible for a considerable share of output. For example, in sub-Saharan Africa, smallholders account for 90 per cent of all agricultural production (Cohen *et al.*, 2008: 31).

There has been some increased instability in cereal output linked to biofuels. In the US high maize prices in 2006 encouraged some farmers to switch from soy and wheat to planting maize in 2007. The US was able to meet all its maize demands including its exports commitments. However, the prices of soy and wheat increased, which was considered likely to cause farmers the following year to switch back to those crops reducing the output of maize (FAO, 2008b). Instability in output feeds through into prices is discussed in the next section.

Food price increases and impacts on the poor

Food prices at the global aggregated level started to increase in 2002 and rose sharply in 2006, 2007 and 2008. 'By mid-2008 real food prices were 64 per cent above their 2002 levels' (FAO, 2008a: 9), despite the good harvests reported in 2008 (FAO, 2009). Although prices had fallen back somewhat by the end of 2008, they were still 17 per cent higher than two years earlier (FAO, 2009). The link with the demand for biofuels was made when, during the first three months of 2008, the price of vegetable oils (which can be refined into biodiesel) rose 97 per cent over their price in the comparable period in 2007 (FAO, 2008b). The fluctuation in the price of food commodities on the world market is not unusual but what raised concern was the simultaneous increase in so many commodities.

The above description of fluctuations in food prices is framed in terms of global averages, and food prices show considerable variation at the national level. How prices at the international level are experienced by individual households is complex and depends on a number of mediating factors, including governmental policies, national levels of production, levels of imports, food stock levels and food aid. Variations also occur between countries (exporting/importing), within countries (rural versus urban areas), and in commodity types (for example not all food crops are traded on the world market). Some governments, such as India and Philippines, intervene using a variety of instruments to protect consumers from large price increases in staples, whereas other governments, such as China and Thailand, allow the market to determine prices (FAO, 2008b). Nevertheless, the concern about food security in poor households when staple food prices rise is linked to evidence that for these households more than half of their total expenditure is on food. There is also evidence to show that when food prices rise, mothers will reduce their own calorific intake in order to secure their children's nutrition which has implications for meeting the MDGs (FAO, 2008a).

Some countries have benefited from the high global food prices but others have suffered, especially net importers of major cereals (rice, maize and wheat) that form staple foods. For example, maize is an essential staple in Latin America and many African countries, where it is consumed by more than 1.2 billion people (CFC, 2007: 36). The price of maize has increased significantly since the turn of the century. In March 2009 the price of maize in Mozambique in US dollars was 29 per cent higher than a year earlier and in Kenya it was 43 per cent higher. In Nicaragua, a 45 per cent increase in maize prices between March 2008 and March 2009 has been reported and in Guatemala 35 per cent in the same period (FAO, 2009). Maize provides around 40 per cent of the average daily calorific intake in Guatemala (Christian Aid, 2009). If a staple becomes expensive, people substitute their traditional staple with a cheaper one. However, if all staples simultaneously increase in price the risk becomes a reduction in calorific intake, because people's coping strategy of switching to cheaper alternatives is disrupted.

The situation in terms of food insecurity is worse in countries with existing high levels of chronic hunger, and that are also net importers of petroleum products, since the price rises of these products have an impact on the transport and storage costs of food. However, in countries where non-internationally traded crops, such as cassava and sorghum, are the food staples households have not been affected by the international food markets. The countries in this group are often among the world's poorest countries.

Food price rises are felt most by people who cannot produce or buy enough to feed themselves adequately. In the first analysis this would obviously point to urban households; but it may be somewhat surprising to find that many rural households, including those owning some land, are net purchasers of food. Indeed FAO considers that the majority of poor rural households are net food buyers. Single harvests and lack of storage facilities create what is known as the 'hunger season' when household food stocks run out forcing them to buy food on the open market. FAO has also found a gender bias in the impacts of high food prices on household welfare. Woman-headed households fare worse than comparable man-headed households, which is attributed to woman-headed households spending a higher proportion of their income on food (FAO, 2008a). Food price rises have a significant impact on poor households who use up to 40 per cent of their budgets on food (FAO, 2009). Households are left with difficult choices: divert funds from other necessities, buy smaller quantities of staples or substitute for cheaper (often nutritionally poorer) food.

Poor people in low income food deficit countries are particularly vulnerable to food insecurity. To bridge deficits food is bought on the world market at high prices reducing the quantities the poor can afford to buy. In addition the situation is exacerbated if poor households are reliant on food aid. The aid agencies have fixed budgets and so when world market prices increase they purchase less food and have reduced stocks to distribute (Clements, 2008; Eide, 2008: 13). The data available in the Food Aid Information System of the World Food Programme shows that for maize, the staple which showed the largest price increase between 2006 and 2008, food aid deliveries dropped dramatically between 2005 and 2007: 973,458 tons in 2005, 896,258 tons in 2006 and 514,101 tons in 2007.[6] In 2008, when food prices reached their peak, the World Food Programme had to launch a special appeal to donors for US$755 million to cover the additional costs generated by the increases in commodity and fuel prices (World Food Programme, 2009: 4).

There can also be hunger in the midst of plenty. In the US, 0.5 per cent of households are classified as suffering from chronic hunger (Paarlberg, 2010). The Food Corporation of India reported that in 2002 food grain stocks in India were at an all-time high of just over 58 million tonnes against an annual requirement of around 10 million tonnes for ensuring food security (Food Corporation of India, 2003). However, newspaper reports state that, at the same time in India, an estimated 200 million people were underfed and 50 million were on the brink of starvation (Goyal, 2002).

Are biofuels to blame for food price rises?

There are a number of factors other than the increase in biofuels output which have been identified as contributing to food price rises. These factors include the rise in oil prices, the increase in food demand by emerging economies, increasing costs of agricultural production, the depreciation of the US dollar, financial markets speculation with commodities, hoarding along all parts of the production chain, as well as drought and poor crop harvests in some producing countries, such as Australia.

The price of oil, which peaked at US$145/barrel in July 2008, has had impacts throughout the food supply chain. The cost of artificial fertilisers and transport were particularly affected. The cost of some fertilisers tripled in the first two months of 2008, while freight transport costs doubled in the 12 months after February 2007.

In 2006 there was a decrease of 60 million metric tonnes of cereal output linked to the weather in the main producing countries of North America, Europe and Australia (CFC, 2007). Such a level of reduction in cereal output would be expected to result in price increases. One of the traditional response mechanisms to stabilise prices is for governments to release part of their buffer stocks onto the market. However, the steady global food output for around 20 years had induced policy changes by governments to reduce the level of food held in store. From the early 1980s, the ratio of stocks to use has fallen from more than 30 per cent to around 15 per cent. This leaves international cereals markets vulnerable to relatively minor variations in availability, which translates into price rises (Wiggins *et al.*, 2009). The private sector made similar decisions to keep their costs down. Both public and private sector actors had been able to make use of trade liberalisation to buy cereals and grains as and when they required, reducing the need for them to maintain food stores (Trostle, 2008). As a consequence there were insufficient stocks of food to respond to the situation that occurred in 2008.

Economic development and income growth in developing and emerging countries, as well as population growth and urbanisation, have been gradually leading to a change in diets. There is increased demand for more meat and dairy products, which is intensifying the demand for feed grains.[7] Approximately 30 per cent of world's grain supplies are reported to be used as animal feed (UN–Energy, 2007: 33).

The real currency exchange rate also plays a role in food prices. The US dollar is the currency used to trade many crops internationally, and the relatively weak dollar against many currencies in the first decade of the 21st century[8] has largely lessened the impact of price increases; but it has also increased the demand for grain imports from the US (FAO, 2008b: 75; Heady and Fan, 2008). On the other hand, the relatively high Brazilian exchange rate in 2008 has slowed the expansion of soy (Searchinger, 2009).

Food-exporting countries have contributed to price rises on international markets by placing bans on exports to protect their own consumers (e.g. Egypt, Vietnam, Cambodia and Indonesia banned rice exports). At the same time, importing countries, afraid that prices would escalate even further began to buy extra stocks. This brought a similar reaction along the chain down to the level of consumers who, probably driven by incorrect media reports of food shortages, took to panic buying (Paarlberg, 2010). Exporters and importers entered into a vicious circle.

In 2006 there were also new entrants into the world agricultural commodity markets in the form of hedge funds, index funds and sovereign wealth funds looking to diversify their portfolios. With so many contributing factors affecting the prices of commodities, it is difficult to attribute the exact influence of these actors; indeed economists and financial analysts seem unable to agree on whether or not these new actors had any influence (see Baffes and Haniotis, 2010, for a summary of the arguments). However, it has been suggested that the way such funds trade and manage their portfolios contributed to short-term price volatility of some commodities by influencing the decisions of farmers, traders and processors of agricultural commodities (FAO, 2008b; Trostle, 2008). Also the sheer size of the funds[9] from this new source relative to the size of commodities markets might have helped push up food prices (Baffes and Haniotis, 2010).

However, to what extent has the increased production of biofuels played a role?

Several international development agencies, including the World Bank and the International Monetary Fund agree that the increasing demand for feedstock for the production of biofuels has played an important role in the rise in food prices (IMF, 2008, and World Bank, 2008b, and quoted in Evans, 2009: 14). However, there is less agreement about the extent of the contribution (Pfuderer *et al.*, 2010). This can be explained by the fact that they were using different models and scenarios with different assumptions and which factors are included, hence they get different answers. Indeed, one economist for the Asian Development Bank considered that it was 'near impossible' to quantify contributions based on models (Timmer, 2008, quoted in Pfuderer *et al.*, 2010: 38). There is a consensus that the increasing demand for feedstocks in the US and the EU for biofuel production had a direct impact on the rise of maize and oilseeds prices, and indirectly on wheat and soybean prices by means of induced land-use changes to produce biofuels feedstock instead of food. In a paper for the World Bank, Mitchell attributes a share of 70–75 per cent to biofuels in the increase of food commodities prices (Mitchell, 2008). The IMF calculates a share of 60–70 per cent in the rise in maize prices and 40 per cent with respect to soybeans prices (Collins, 2008, and Lipsky, 2008, quoted in Mitchell, 2008: 4; and Heady and Fan, 2008: 8). A study for the International Food Policy Research Institute (IFPRI) (Rosegrant *et al.*, 2008, quoted in Mitchell, 2008: 4; Leturque and Wiggins, 2009: 1) estimates that between 2000 and 2007 increased biofuel demand accounted for 39, 21 and 22 per cent of the increase in the real prices of maize, rice and wheat respectively. Another study by the IFPRI analysed different factors that were considered possible drivers of commodity price rises. This study concluded that increased production of biofuels and increased oil prices offer the most convincing explanations for increasing prices across different commodities, and biofuel production providing the strongest explanation for the rise in maize prices (Heady and Fan, 2008). The divergence in conclusions between all of these studies demonstrates the difficulties in establishing the extent of influence of biofuels (or any other factor) on food prices. Indeed, the FAO considers that it is not possible to quantify accurately the contribution of biofuels demand to increases in commodity prices.

Between 1980 and 2002 there was a steady increase in demand for wheat and what are known as coarse grains (maize, barley, sorghum, rye and oats). Some of this demand was for fuel ethanol (7 per cent of increase) but the vast majority was for feed use (44 per cent), and food and other non-feed use (49 per cent excluding US ethanol) (Trostle, 2008). What appears to have been the significant impact on global grain prices was that between 2002 and 2007, an additional 53 million metric tonnes of US maize was used to produce ethanol which accounted for 30 per cent of the global growth in wheat and feed grains use (Trostle, 2008). However, Paarlberg, a leading authority on food policy, considers that biofuels demand was not the leading driver because maize prices would have risen more sharply than wheat or rice (neither of which are used extensively as biofuel feedstocks), while in fact maize prices rose less sharply than the other two grains (Paarlberg, 2010).

There has been a long-term trend in a slow increase in food prices. FAO considers that a significant demand for biofuels, together with high oil prices that contribute to the increasing cost of food production, will be a factor contributing to high food prices becoming a permanent reality. This forecast is based on the assumption that the demand for first generation biofuel feedstock is likely to continue growing rapidly due to high oil prices and governmental policies supporting substitution for liquid transport fuels. The IEA estimates that the share of the world's arable land devoted to growing biomass for liquid biofuels could increase from 1 per cent in 2004 to between 2.5 and 3.8 per cent in 2030 (FAO, 2008a: 21). The estimates are based on the assumption that liquid biofuels will be produced using conventional crops, that is, first generation feedstocks. Therefore, based on this analysis, it is the first generation biofuel feedstocks derived from edible crops which are likely to pose the greatest threat to food security when they compete for prime agricultural land. However, if investment in research and development to bring second and third generation technologies to maturity continues, then these options potentially reduce the competition with food crops. Alternatively, the floor could drop out of the biofuels market if hydrogen or electric vehicles become the promoted options for transport fuels. This would result in a waste of investment by Southern economies, unless biofuel developers target domestic as well as export markets.

Biofuel feedstock plantations, land and food security for the rural poor

The IEA forecast that if biofuels are to provide 26 per cent (in terms of energy content) of transport fuels by 2050 they will require 160Mha of crop land (IEA, 2008, quoted in UNEP, 2009: 66). If correct in terms of the scale of crop production required to satisfy the increasing global demand for biofuels, then such a level of output may bring significant land-use changes. These changes can be social, in terms of land ownership, which determines what the land produces and who has access to the land and its products, and ecological, which determines the biodiversity of the land. As was discussed in Chapter 4, biodiversity changes when non-agricultural land is converted to crop land or the nature of the agricultural systems changes, for example, when rangeland is converted to grow biofuel crops or one crop is substituted by another crop more suited to biofuels. This section examines how these changes might affect the food security of the rural poor by looking at how access to land, food availability and income-earning opportunities are affected by biofuel programmes.

Access to land

As described in Chapter 3, when the institutional framework directly or indirectly facilitates or stimulates the establishment of large-scale plantations for biofuel feedstocks, access to land may be threatened through increased concentration of land ownership. Ownership of the land is a key factor influencing decisions about land-use changes, because the owner will determine which crops are grown and what management practices are used. These decisions also affect ecosystems.

Access to land allows individuals and households to grow their food, a key factor identified by Sen for food security at the micro-level. Consequently, jeopardising the food security of those depending on land-based agricultural livelihoods by biofuels projects is to be avoided (Cotula *et al.*, 2008: 14). The other response to be avoided is an increase in the number of landless people who move to the cities increasing the number of the urban poor, where they enter the group of net buyers of food vulnerable to high prices. Tracing this effect solely to an expansion of biofuels is not easy.

The IFRI estimates that in the period 2006 to 2009, between 15 and 20Mha of farm lands in the South had changed hands. While not all of this land is intended for biofuels, a concern is that it is often the higher value land (that is, land with good rainfall, access to irrigation and proximity to markets) that is being acquired by large-scale investors. In Mali, for example, all recorded land deals are concentrated in the agricultural zones with the highest potential (Vermeulen and Cotula, 2010). Monitoring land deals is difficult because many countries do not have the bureaucratic capacity to track such transactions. Researchers often have to rely on the media for information which are found to have conflicting reports (Von Braun and Meinzen-Dick, 2009).

The acquisition of land occurs through a number of mechanisms both legal (direct purchase from smallholders or termination of leases of tenant farmers) and illegal (violation of land rights by not recognising customary law or land seizure, in some cases by violent means) (Cotula *et al.*, 2008: 32). The literature abounds with examples of smallholders relinquishing their land (see Chapter 3 for a discussion of why this is happening and Box 5.2 for some examples). However, the literature is less clear on quantifying the exact impacts these changes have on access to land. A study in Tanzania has highlighted examples of good and bad practices by companies acquiring land for growing biofuel crops (Sulle and Nelson, 2009). It appears that decisions about land allocation do not always take account of the implications of land-use changes on villagers' livelihoods. When large tracts of land are acquired by biofuel companies, it is not always land under agricultural production that is bought but land that can be considered common property, such as miombo woodland. So it appears that agricultural crops are not being displaced; but instead a reduction in access to other important household goods and services, e.g. commercial charcoal production and harvesting products such as traditional medicines, mushrooms, fuelwood and building materials, is occurring. It has been estimated that informal and non-industrial uses of forests in Tanzania add approximately US$35–50 to national annual per capita income (Sulle and Nelson, 2009). In other words, household ability to buy food is being undermined.

Gender also plays a role, as farming contracts are often with the male head of household. This has two possible consequences in relation to household fuel security. First, based on experiences with commercialisation of agriculture, there is fear that land normally managed[10] by women, which is used for producing crops and hence forms the basis of household food security, can be appropriated by men for producing biofuel crops. Second, there is a substantial body of evidence showing that within the households it is women's rather than men's cash income that is used for buying food. Therefore, who

---

Box 5.2 Changing land rights and biofuels

In Guatemala, there are reports about the displacement of tenant farmers who used small parcels of larger farms to cultivate food for household use. This change has been linked to a process of land consolidation where large plantations are integrated into even larger landholdings for the production of biofuels in Fray Bartolomé Las Casas in Alta Verapaz, as former landowners are reorganising their farms to sell them to biofuels investors.

Source: Hansen-Kuhn, 2008.

In the Wami Basin, in Tanzania, a thousand rice farmers may be evicted as a result of a sugarcane plantation.

Source: Cotula *et al.*, 2008.

earns cash from biofuels can affect whether or not household food security is improved. The approach taken by the Hassan Biofuels Park (Karnataka, India) addresses this aspect, since women are encouraged to plant biofuel crops in their backyards where they already have tree crops (Narayanaswamy *et al.*, 2009). Anecdotal evidence would suggest that these women intend to use at least part of their income for buying food.

On the positive side, a novel method of enabling access to land for rural landless people has been tested in Ranga Reddy District in the Andhra Pradesh, India. Where land ownership is not clear or where ownership can be claimed through use, allowing landless people access to land can be resisted. However, by granting landless people usufruct rights on non-forest, low quality private and common property land near their villages avoids conflict over ownership while allowing very poor people the opportunity to earn an income (ICRISAT, 2007).

Impact of land-use changes on food availability

Plantation schemes for biofuel production can affect local food security through land-use changes in three ways:

1   direct displacement of food crops
2   expansion into uncultivated regions
3   use of waste land.

The spread of any monoculture crop displaces the production of other food crops affecting local availability of staple foods and subsistence crops, the latter provide the buffer for household food security in periods of vulnerability (Rahman Osmani, 2010). The gap can be filled by importing food from other regions of the country, although it is likely to be at higher prices since transport costs may be significant. For example, in Colombia, Valle del Cauca has been the main region for sugarcane-based ethanol production where approximately 81 per cent of the country's sugarcane is cultivated and three of the country's five ethanol distilleries are located (Sistema Nacional de Competitividad, 2009). Large-scale cultivation of sugarcane in Valle del Cauca started at the beginning of the 20th century and expanded between the 1950s and 1960s. Since that time sugarcane monoculture has dominated agricultural land use in the region: 50 per cent of agricultural land is dedicated to sugarcane plantations, 20 per cent to other permanent cash crops (coffee and cacao) and 30 per cent is for food crops (Secretaría de Agricultura y Pesca del Valle del Cauca, 2008a). As a consequence around 81 per cent of the food that enters the main food stock and supply centre of Valle del Cauca is imported from other regions (Secretaría de Agricultura y Pesca del Valle del Cauca, 2008b). In Brazil, the National Supplies Company (Companhia Nacional de Abastecimento – CONAB)[11] admitted that in 2007 the areas cultivated with maize, soy and wheat were decreasing in the states of Mato Grosso, Minas Gerais, São Paulo and Paraná – where the largest number of sugarcane plantations are located and more than 70 per cent of Brazil's ethanol is produced (Walter *et al.*, 2008) with possible adverse effects on food production and food prices (Martins de Carvalho, 2007, quoted in Monsalve *et al.*, 2008: 55). Furthermore in 2007, in the sugarcane-producing regions, the area dedicated to the cultivation of beans – a staple of the Brazilian diet – decreased by 261,000ha, which represents 12 per cent of the total production of beans in the country (Hansen-Kuhn, 2008: 14).

Land-use changes that occur when forest, woodlands and savannahs are converted into biofuel production systems can deprive the rural poor, especially landless people and forest dwellers, of natural resources that form the basis of their livelihood strategies, including providing essential components of their diet as well as income generation opportunities. The Forestry Department of Senegal estimates that creation of jatropha plots at the expense of forest in the Bignona area could represent a decline of 68 per cent in income sources for rural people, as well as affecting the availability of nuts and other forest products used to supplement household nutritional needs (Hansen-Kuhn, 2008). These natural resources provide a safety net, particularly for poor households, in periods of hunger, e.g. in the season preceding crops harvests, when crops are attacked by pests or diseases, when there is drought or when harvests have been insufficient (Falconer and Arnold, 1991; Guijt *et al.*, 1995: 7).

There are also indirect effects on food availability caused by large-scale biofuel projects, due to the demand on groundwater and surface water supplies (see Chapter 4).

The use of the so-called 'idle', 'degraded' or 'marginal land' for biofuels plantations is being considered by some governments as a response to concerned voices regarding competition between fuel and food for land, as well as the environmental consequences of clearing forest or land with high biodiversity (see Chapter 4). Some of the candidate crops for biodiesel (e.g. jatropha) can be grown on marginal lands. India is planning to use of 17.4Mha of wasteland for jatropha production (TERI, 2004). The risk, however, is that what is considered idle, marginal or degraded land by governments and investors can be viewed very differently by the community where this land is located. Often it is this common land which plays an important role in the livelihoods and food security of the rural poor (Cotula *et al.*, 2008: 22–3; Gaia Foundation *et al.*, 2008; IUCN/DFID, no date). A survey in Hassan district, Karnataka state, India, found that farmers and the government had distinctly different perspectives on the way to classify land (Narayanaswamy, 2009). Land that fell under the official classification of waste land was viewed by farmers as an important part of the farming system as supplementary grazing land, and other community members had multiple uses for the land including gathering fuelwood, medicines and flowers for religious purposes.

Women are considered the most vulnerable to changes in marginal land use because, due to the significant gender inequalities in land ownership in developing countries, they rely the most on its resources to provide services (see Chapter 2). Evidence from India, West Africa and sub-Saharan Africa demonstrate that land classified as marginal is traditionally used by women to grow crops and for harvesting natural resources for household consumption (FAO, 2008b; Rossi and Lambrou, 2008). According to the FAO, female-headed households belong to the more vulnerable group of people to the increase in food prices (FAO 2008a), therefore the use of marginal land for biofuel production may worsen their food insecurity status.

Several examples of the negative impacts of shifting 'marginal', 'degraded' or 'idle' land into biofuels feedstock plantations have been reported. Hansen-Kuhn (2008) cites two examples from Ghana. In Alipe village in the north of the country, external investors were planning to establish jatropha plantations for biodiesel, which entailed the destruction of the sheanut trees growing on the proposed plantation area. The sheanuts are used as food when harvests run out as well as a major source of income for women during the rainy season (see Box 5.3). Villagers concerned about the loss of the trees protested and were able to stop the project (Hansen-Kuhn, 2008). In Makango, a small town on the Volta River, the surrounding marsh lands were considered by biofuel investors as suitable for sugarcane for ethanol production. However, these marsh lands support thriving fishing communities' livelihoods which would be threatened by the draining of the marsh lands.

In Indonesia the government allows oil palm plantations on lands classified as 'marginal', 'critical' and 'sleeping', despite this land providing food and income for shifting-cultivation farmers, hunters and indigenous people. The conflict between rural dwellers and palm oil plantation companies in Sanggau District in West Kalimantan is well documented. Rural people consider that their livelihoods are being undermined by the expansion of palm plantations into land they use as a source of ecosystem services. They have successfully used a variety of tactics to halt the expansion or at least gain acceptable compensation from the companies (Cotula et al., 2008).

On the other hand, biofuels have been replacing other non-food cash crops such as tobacco in Malawi (Peskett et al., 2007) and cotton in Mali (Cotula et al., 2008). The transition had been stimulated by falling prices of these commodities on the world market leading to a decline in income for

---

Box 5.3 Lack of consultation

'Look at all the sheanut trees you have cut down already and considering the fact that the nuts that I collect in a year give me cloth for the year and also a little capital. I can invest my petty income in the form of a ram and sometimes in a good year, I can buy a cow. Now you have destroyed the trees and you are promising me something you do not want to commit yourself to. Where then do you want me to go? What do you want me to do?'

Woman villager confronting biofuels developer in village of Alipe, White Volta River, Ghana. Source: http://biofuelwatch.org.uk/docs/biofuels_ghana.pdf, (accessed 20 September 2010).

small-scale farmers. However, one needs to be sure that this does not lead to ILUC if production switches to elsewhere.

Increased income from employment

As was discussed in Chapter 3, biofuel production based on plantation-style large estates are promoted for the potential contribution that they can make improving rural employment. It is rather surprising that such claims are not more tempered because past experiences from other agro-chains with plantation modes of production have shown that there is the risk of a potential trade-off between land access and employment. For example, when sugarcane was introduced in southern Bukidnon province of Philippines, many households lost their access to land due to the establishment of large-scale sugar estates. This transition was found to have contributed to income inequality since there was not a net increase in demand for labour (Bouis and Haddad, 1994 quoted in FAO, 2008b: 83).

However, the evidence to date in relation to biofuels is that there is a trend, at least in sugarcane cultivation, to reduce labour through mechanisation (see Chapter 3). The jobs created tend to be seasonal, unskilled and lowly paid, particularly for women. Although, at least in Brazil, the wages for labourers on sugar estates are higher than those in comparable jobs for other crops (Kojima and Johnson, 2005).

Biofuel plantations may not guarantee the same level of employment and food supply that previous access and use of land provided, for example when they replace other cash crops that are more labour intensive. In the Magdalena region of Colombia oil palms have replaced bananas with a corresponding reduction in levels of employment: banana plantations required 1.5 employees per hectare while palm oil plantations require only one employee for every 10ha. The change in crop has been linked to a decrease in food security because bananas were an important part of the diet in Magdalena and local output has diminished (Goebertus, 2008). Although bananas could be imported from other regions, they would probably be more expensive.

In the district of Kisarawe, Tanzania, with a population around 11,000 people, a company planning a jatropha plantation estimated that initially about 1,500 jobs would be created for clearing the land and in the long term there would be employment for approximately 4,000 people (Bailey, 2008: 22). At the present level of technology, jatropha harvesting is not highly mechanised but there are development efforts underway for mechanical harvesters, which would be used to reduce labour costs. Therefore, in the future, the level of employment may be reduced.

In other projects, communities have not been given sufficient information to make informed choices. For example, in the case of the marsh lands in Makango, Ghana, the community was not informed about the nature or the number of jobs that the project would generate. Hence they did not know whether or not the loss of their livelihoods derived from the marsh lands would be compensated by new employment opportunities (Hansen-Kuhn, 2008).

There is more positive evidence of increase in incomes when farmers are employed as outgrowers and with models that are constructed to actively incorporate smallholders and the landless into production chains. Mali Biocarburant SA is a private company producing biodiesel from jatropha and includes the smallholders growing the jatropha among its shareholders. The additional income for smallholders from selling the nuts is estimated at West African CFA1,250/day (€1.90/day) compared to current alternative sources of income of a maximum of €1.15/day. It is anticipated that the farmers will also receive additional income from dividends (Hetterschijt, 2009; Mali Biocarburant SA, 2009).

Pro-food security biofuel production

The question is: can biofuel programmes be designed so that they do not undermine food security? There are clearly examples in the literature which show that this is possible and in some cases have actually enhanced food security. (See for example ICRISAT, 2007: 17; FAO, 2008a: 81; Haralambous *et al.*, 2009; Karlsson and Banda, 2009; Narayanaswamy *et al.*, 2009; Practical Action Consulting, 2009). However, there is little comparative crop yield data provided in the literature to substantiate the claims.

Given that most rural households are net purchasers of food, their capacity to do so can be enhanced when biofuels become a new source of household income. Therefore increased household income enables potential access to better food.

There are a range of different models reported in the literature that promote either small-scale production for self-consumption or integration of smallholders into large- and medium-scale production systems (see Box 5.4). These schemes seem to rely generally on non-edible oilseeds, with the notable exception of sweet sorghum for ethanol production as promoted by the ICRISAT for use in grain sorghum growing areas. Sweet sorghum has the potential to promote food security by producing grains for household consumption, income from selling the juice for fermentation, and the residues for feeding cattle (ICRISAT, 2007).

The institutional arrangements seem to be the key to ensuring food security and pro-poor outcomes with biofuels. NGOs, CBOs and development institutions have been acting as catalysts in the configuration of partnership between small-scale farmers, large processors and producers of biofuels, governmental agencies and funding institutions. Such organisations are able to mediate to ensure the fairness of and compliance with agreements. They also provide training, agricultural inputs and funding, assets which the rural poor generally do not have access to. Food security may be also enhanced by the promotion of intercropping food crops with energy crops. The yield of food crops could benefit from inputs, such as fertilisers, supplied for the energy crops. Soil fertility is promoted by the use of organic fertilisers that are by-products of the oil extraction process, such as seed cake in the case of jatropha (Practical Action Consulting, 2009) or by intercropping with

Box 5.4 Different models for combined biofuel production with increased income and enhanced food security for poor households.

Self-consumption

Women in Gbimsi, a small town in Northern Ghana, grow jatropha on 4ha of land. The extracted oil is used to power shea butter processing equipment. This equipment helps reduce postharvest losses and improves incomes.

Source: Karlsson and Banda, 2009.

The village of Powerguda in Adilabad district of Andhra Pradesh State, India, is participating in a project to produce biodiesel oil from locally grown seeds of pongamia, neem and other plants. The oil is used locally and sold on the market. Women earn about four US$0.04/kg of pongamia seed crushed. The oil extractor crushes about 50kg seed per hour and can run on the biodiesel that it produces. The women also sell the press cake (residue remaining after oil extraction) for use as a soil improver to farmers for about US$0.10/kg.

Source: ICRISAT, 2007.

In another tribal village, Chalpadi, in Andhra Pradesh women are using straight pongamia oil for running a 7.5KVA generator to provide electricity to light their homes. It takes 5–6 litres of pongamia oil to produce 10–12kWh of electricity.

Source: ICRISAT, 2007.

Using intercropping

Mali Biocarburant promotes intercropping of jatropha with other crops in its biofuel schemes in Mali and Burkina Faso. It has carried out field research on optimal spacing for intercropping and provides extension work for participating farmers. Although 80 per cent of farmers began with a monocrop, it is claimed that 87 per cent now practice intercropping.

Source: Hetterschijt, 2009.

Using non-agricultural land

In Guatemala, under the Biodiesel for Rural Development project, unused land was allocated by the government to small-scale farmers organised in cooperatives or clusters to plant jatropha and sell the oil to larger processors for biodiesel production. The project area was selected on the basis that there was no competition for the land with food crops and the system of live fences using jatropha was already part of the agricultural system.

Source: Practical Action Consulting, 2009.

The Biofuel Park programme, Karnataka, India, encourages rural people to grow non-edible indigenous oil on bunds, as hedges and in backyards. These plots of land are not used for agriculture, which avoids interfering with food production; nor are they on common land that provides ecosystem services. The land used is close to where farmers live and work. Women particularly like being able to use their backyards since they can combine care of the plants with their regular household chores. The programme aims to provide a range of species that yield seeds at various times throughout the year, thus ensuring an even oil output and stable income throughout the year.

Source: Narayanaswamy et al., 2009

leguminous crops (Vaidyanathan, 2009). The success of neem and *karanj* seed-cake as a bio-pesticide has also been reported (Vaidyanathan, 2009). If the by-products can replace artificial fertilisers and pesticides and either enhance or at least do not reduce yields, then this can potentially decrease household expenditure on chemicals, freeing cash which could be used for buying food. Farmers' cooperatives in southern Brazil, having food security as one of the aims of growing biofuel, promote a model of intercropping, using both tree and food crops, in which each producer is only allowed to plant 2ha of biofuel crops (Wilkinson and Herrera, 2008, quoted in McMichael, 2010: 620).

Schemes that recognise and respect traditional land holding systems are less likely to damage food security. There are examples in the literature where companies have made efforts to respect local land holding systems and to integrate smallholders into production chains as outgrowers (see for example Cotula *et al.*, 2008; ICRISAT, 2007). However, it is too early to assess the impact such schemes have on household food security.

The literature does provide some evidence to show that it is possible to produce biofuels, even on a large-scale, without jeopardising the rural poor's food security. Important components in ensuring food security are the institutional arrangements and enabling regulatory frameworks which ensure land rights and stimulate partnerships between small-scale farmers and private companies/investors. The involvement of NGOs, CBOs and development institutions has been instrumental in promoting better land management practices and for assisting the rural poor in obtaining fairer treatment from biofuel developers.

Conclusions

The debate about the competition between food and fuel for agricultural land is not new dating back at least to the 1970s in the wake of the first oil price shock when countries began to consider energy crops as an alternative to petroleum fuels (Rathmann *et al.*, 2010).

In this chapter, we have assessed the evidence about whether or not biofuels have undermined food security. We have used two approaches to do this assessment. First, the framework that FAO and other international development agencies use to determine food security: availability, stability of supply, access, which is related to price, and nutritional value. The first three operate at both the macro- and micro-levels while the fourth is at local level, i.e. household, effect. Second, we drew on the work of Sen who showed that what people need to ensure food security is the means to produce or buy food.

In terms of availability, at the global aggregate level, there was in 2010 enough food in the world to feed the global population. Certainly at least for the short term, the issue of availability is not a cause for concern. The underlying problem is one of distribution. First, in the physical distribution of food: although the quantity of food produced worldwide as of 2010 was sufficient to feed everyone, there are still hungry people in the world. In 2009, FAO considered that 1,020 million people were undernourished

(Paarlberg, 2010: 35). Second, it is a distribution problem of the type of assets Sen regarded as the necessary basis of food security: income to buy or to have access to land to produce food. Biofuels have been promoted for the influence that growing and processing these crops can have on the former, which is significant because over half of the world's hungry are considered to live in smallholder farming households who could be a target group for inclusion in biofuel value chains (BVCs). In respect of access to land, there are concerns that growing biofuels is having an impact on land ownership and on the way that land is used.

It is difficult to demonstrate causality between growing biofuels and negative impacts on food availability, in part because the crops used at present to produce biofuels are able to switch between different food production chains. Indeed for many large-scale producers, this ability to switch between chains to get the best available commodity price is one of the main attractions. The available evidence shows that there has been some displacement of *commercial* food crops, in particular wheat and soy, but there appears to be no significant switching of smallholders from subsistence crops or crops intended for the local market. Indeed there are claims (but little empirical data) that food availability has been enhanced by biofuel projects that have focused on incorporating better land management for smallholders, for example, by promoting intercropping and the use of fertilisers from residues.

Much of the anger about biofuels has been directed at US maize being used to produce fuel ethanol. However, in 2008, the US was still able to meet its maize export commitments so the impact of biofuels here is related not to food availability on the world food market but to food prices (see below in this chapter). There is evidence from Brazil that coffee, oranges and soy have been displaced in Paraná and Saõ Paulo states, by sugarcane production (Rathmann *et al.*, 2010) and soy production is being displaced on to pasture land, which affects meat production that in turn can be displaced elsewhere. While this has had an impact on respective commodity prices, there is no data about the effects on the food intake of the poor.

Probably of greater concern for the food security of rural households has been the land-use changes related to the use of so-called 'waste or non-productive land' for biofuel production. This land provides an important source of food for the vulnerable, and at times of vulnerability for a larger portion of the rural population. It is also an important source of saleable commodities that make a significant contribution to rural households' income. The rural poor, particularly the landless, are having their capacity to produce food or earn the money to buy food reduced if they lose access to land an outsider considers non-productive. The irony of this situation is that the promotion of 'waste land' for biofuels has been in part in response to the criticisms of biofuels undermining food security. The use of waste and non-productive land has been particularly promoted in Africa and India where food insecurity is notably problematic.

Biofuels have had an effect on land availability for food crop production by influencing the value of land. A study carried out in Paraná state, Brazil, reported in Rathmann *et al.* (2010), shows a correlation between the increase in sugarcane planted areas between 1995 and 2005 and an accentuated rise in land values in the same period. This effect has been blamed for a shift towards more marginal land, despite the need for more inputs such as fertilisers to make such land productive, which can increase input costs and reduce profit margins. Increasing land prices can also put pressure on smallholders to sell their land. In Chapter 3, we reported on the decline in the number of small-holders in Brazil.

The first decade of the 21st century saw an unprecedented rise in food prices in the global marketplace. These price rises had a major impact on the poor, particularly the urban poor and those in receipt of food aid. There were protests about food prices across the globe in 2007 and 2008 not only from consumers but also from people whose livelihoods are connected to food production (e.g. millers and bakers in Malaysia, market traders selling soybeans and meats in Indonesia and wheat marketers in Pakistan) (Trostle, 2008). There is no doubt that biofuels played a role in these high food price rises although the exact contribution, since there were a number of factors, is still hotly debated. However, biofuels were not the only factor and there is no consensus about how much biofuels contributed to these price rises. Also the growth in biofuel production cannot be blamed for the price increase of all food crops, e.g. rice which is not used for bioethanol (Wiggins *et al.*, 2009) nor have all staple food crops been influenced by events on the world market because they are non-traded on e.g. sorghum. The degree of transmission of international prices into national markets depends on different mediating factors at the local level, such as dietary patterns or livelihood strategies, or government policy. Countries have not all experienced the impacts of the food price spike of 2008 in the same way.

The situation in respect of food prices in 2008 has been called by World Bank President Robert Zoellick[12] 'a perfect storm',[13] when a number of both short-term effects (such as droughts) and long-term effects (such as decline in the relative growth of agricultural output to population growth) happened simultaneously. It should not be forgotten that other factors, such as population growth, changing diets and climate change, can also affect food security. Economists predict that, in the longer term, adjustments in food availability will occur as farmers respond to the market, although the underlying long-term trend of increasing food prices is considered likely to continue (Wiggins *et al.*, 2009). However, agribusiness farmers serve global markets and are linked to these markets via global information systems. Based on information and stimuli from governments in the form of subsidies and mandates, these farmers are able to switch from one crop to another in consecutive years. Farmers' expectations of the size of the return on their crops can lead to instability of supply which undermines food security. Maize output in the US was certainly a factor but not the only one contributing to the instability.

The events of 2008 on the world food markets appear to have been what economists refer to as a 'spike', that is a dramatic rise and fall in prices over a short time period (in this case a couple of months). Indeed the fall was much stronger than some commentators had expected (see for example Wiggins *et al.*, 2009). Many economists and financial analysts place their faith in markets to eventually resolve the instability in food prices on the world market (Baffes and Haniotis, 2010). However, people 'live in the short term' and from the perspective of the poor, these food price increases have been very painful (Searchinger, 2009). Despite prices falling in 2008, the global economic crisis has made it difficult for the poor to benefit from 'cheaper' food. The effect of the economic crisis on food security includes the loss of income through unemployment, particularly for the urban poor, and a fall in remittances from abroad, on which many low income families in the South depend (FAO, 2009).[14]

Biofuels have had a major effect on food prices, especially maize. As was pointed out above, the US met its export commitments. However, what has received less attention is what effects do US maize exports have on local producers in the South? US maize is exported to many countries that are not food deficient, e.g. Mexico and South Africa. However, US maize is highly subsidised and many local producers in the recipient countries are unable to compete; but there are signs that in Africa, at least, local maize production has been stimulated by the more competitive price of imported maize.[15] There are also signs that smallholders are switching out of export crops to food crops for their own consumption using traditional methods of low input farming, although this would appear to be driven by the high price of fertilisers and petroleum fuels.[16]

In retrospect, now that more data is available and there has been time for more thoughtful analysis, it would appear that biofuels have unfairly borne much (sometimes it feels like all) of the blame for the dramatic food price rises in 2008. This blame has been attributed without understanding the way in which the data on food prices and the factors influencing those prices is arrived at. We would agree with the following:

> In summary, there are many studies with very different estimates of the impact of biofuels on crop and global food prices. Various assumptions are being made to these estimates – some of which are economically reasonable and feasible, while others seem a bit unrealistic. Unfortunately, most media attention has been given to the extreme estimates without much consideration for the economic assumptions that must be made to generate these results.
>
> (Baier *et al.*, 2009, quoted in Pfuderer *et al.*, 2010: 42)

If we look at Sen's other requirement – the capacity to earn income from biofuels – it is unclear whether or not jobs created by the biofuels industry can compensate for any negative influence on food security for the rural poor related to deprivation of access to land and changes in land use by enabling them to buy food (Dufey *et al.*, 2007: 13). Most of the available literature on biofuel production is written by the organisations implementing the projects, therefore, objective data on income is difficult to find. However, the mode of production seems to be significant for the rural poor's income. Smallholders integrated as outgrowers or who own biofuel processing facilities, seem to be able to earn incomes they find satisfactory, if their continued participation in such schemes can be taken as an indicator of satisfaction. There are examples where farmers are clearly not satisfied with the level of remuneration and opt for exclusion (see for example Ariza-Montobbio *et al.*, 2010). Many of the projects focused on smallholders and the landless are opting for non-edible seeds about which there has been little agronomic optimisation. Therefore, there exists the possibility for improvements in income if there is investment in extension work related to these plants. Plantation workers on large-scale estates seem to be the most food insecure. They suffer from low wages and with the increasing likelihood of mechanisation reducing the number of jobs available. Although, this negative impact is not a specific characteristic of biofuels; rather of this mode of production, which is the same used for many other commodities.

A moratorium on biofuel production is not the answer to food security. Likewise, the demand for biofuels is unlikely to go away, even though there has been some reduction in targets, for example, in the EU, partly in response to negative publicity about biofuels and lobbying by NGOs. Indeed if all biofuel production ceased tomorrow it would not solve the problem of food security. Such a moratorium might even be disastrous for many small-scale farmers in the South who have already opted for inclusion in biofuel production chains. These farmers, unlike their counterparts in the North, do not have the capacity to switch crops from season to season. We will explore in Chapter 6 how the production model used in some projects for biofuels that promote energy self-sufficiency at the local level can contribute positively to rural development, directly benefiting the rural poor.

There are ways to ensure that biofuels do not undermine food security: first, in the models of biofuel production promoted; and second, in the direct measures taken to improve food security. Governments can take staples out of the fuel production chain as has happened in some Pacific Island states (Cloin *et al.*, 2007). They can ensure foreign investors respect indigenous land rights. They can also avoid blanket promotion of 'waste' land which is regarded as unproductive. However, it is often not 'waste' – it is just not dedicated to commercial agricultural output. On the contrary, as was outlined in Chapter 4, it is the source of many goods and services which contribute to the rural poor's food security.

There are still possibilities to improve food security by increasing food production with the current levels of land under cultivation through selection of more productive crop species and better land management practices. For example, in India, which is considering a biofuels programme, the average yield of rice is only about half the Asian average and the total area under cultivation in 1990 was approximately the same as it was in 1970. One estimate suggests that the average yield of wheat in India could be increased from 2.6t/ha to 4.0t/ha and maize could be increased from 1.7t/ha to 3.5t/ha (Lal, 2006, quoted in UNEP, 2009: 73). There is underinvestment in improving yields. In Africa, where approximately 60 per cent of the population is dependent on farming, government investment in agriculture is on average around 5 per cent of their budgets. At the same time yields in Africa have been declining and food insecurity increasing (Paarlberg, 2010), although not all of this decline can be attributed to a lack of state investment in agriculture. A cause for concern is the switch from public funded agricultural research and development by state and international institutions to the private sector. This is considered to have possibly contributed to the slowing in growth of improvements in crop yields since the focus of agricultural research has switched to cost-reducing innovations rather than yield-enhancing techno-logical developments (Trostle, 2008). There is certainly a need to give more attention to improving the yields of subsistence crops that are vital to the food security of the poor (Rahman Osmani, 2010).

There is also a need to promote approaches to improving crop yields without the application of expensive chemical fertilisers. For example, inter-cropping with legumes, which fix nitrogen, provide ground cover, control weeds and increase soil phosphorus availability, helps improve soil fertility and soil stability (Lathwell, 1990, quoted in Hart Energy Consulting and CABI, 2010: 9). The legumes can also be part of the household's own food supply or animal feed.

Food availability can be improved by addressing postharvest losses. FAO estimates that about 25 per cent of food produced never reaches the consumer (FAO, 1989). The reasons for postharvest losses are varied, including inef-ficient processing technologies, poor infrastructure, insufficient storage capacity (particularly cold storage which requires a stable electricity supply) and poor transportation infrastructure. Investment in overcoming these weaknesses in the food supply chain would not only improve food avail-ability but also contribute to improving the income of small-scale farmers and agricultural labourers. In Chapter 6, we will explore how energy plays a part in improving these aspects of food availability.

Our conclusion is that the claims that biofuels will starve poor people in the South to feed the cars of households in the North are simplistic and overlook the complexities of food security and influences on food prices. It now appears that biofuels were probably not the most significant influ-ence on the prices of internationally traded food commodities. Instead the entrance of new investors with large sums of money created instability in the markets (one of the warning signals used by international development agencies for undermining food security) and the high oil prices were prob-ably the most influential factors (Baffes and Haniotis, 2010). Adequate food availability at the aggregate level does not necessarily ensure adequate food availability at the household or individual level. Increased household income does not necessarily translate into food security (Rahman Osmani, 2010) nor does an increase in food output necessarily lead to a reduction in hunger. As Sen pointed out, who is able to earn cash income and who controls household resources are key factors in an individual's food security.

6 Do liquid biofuels address rural energy and poverty issues?

This chapter assesses the evidence from the previous chapters as to whether or not biofuels are able to contribute to addressing the energy and poverty issues raised in Chapter 2. Biofuels can potentially play a dual role in addressing energy and poverty issues in rural areas by providing a direct source of income through growing and processing biofuel crops and in their role as an energy carrier. Energy can contribute to combating rural poverty through or providing a means for:

1   improved health
2   increased productivity and new opportunities for additional income
3   reduced manual effort, drudgery and time spent on household survival activities.

The chapter begins with an analysis of how biofuels are able to address issues related to energy poverty, first as an energy carrier at the household level, then at the community level. It goes on to look at how biofuels interact, with some specific issues related to rural poverty identified in Chapters 1 and 2: time poverty and drudgery; food security; ecosystems and biodiversity; gender equity; and income generation. The chapter ends by answering the questions posed in Chapter 2 about the effects of biofuels on rural poverty: to what extent do biofuel production chains really provide an opportunity for the rural poor? What form does participation take? Who is benefiting and who is losing out? How?

Energy poverty

In Chapter 2 the concept of energy poverty was introduced as a dimension of poverty. This concept links the quality of energy carriers, the services people need (such as lighting) and their capacity to pay. Poor rural households live in energy poverty, in other words they use low quality energy carriers and many of the services they need or aspire to are either inefficiently provided or not met at all. Energy poverty, in part, is linked to a lack of capacity to pay but also the limited or non-existent availability of high quality energy carriers in rural areas. Also, within the household, who decides which services will be paid for is a factor determining energy poverty reduction by mediating access to energy carriers. People's capacity to move out of poverty, or resist moving into poverty, is linked to energy poverty because of the impact it can have on health and the capacity to earn income. This section summarises the evidence about biofuels' role in addressing household energy poverty.

The most basic energy services of poor rural households are cooking, boiling water and lighting plus, in some climates, space heating or cooling. Bioethanol could be used as a cooking and lighting fuel at the household level. Bioethanol burns cleaner than fuelwood so there is the potential to reduce indoor air pollution and the associated negative impacts on women's and children's health. However, bioethanol uses a different type of stove to that used with wood fuels, which may require different pots, so the fuel, the stove and any replacement cooking implements have to be available at a price people can afford. The switch to ethanol would require a shift in cooking practice; for example, the method for controlling the heat output of the stove. Such changes can act as a significant barrier to the transition from wood fuels, and are a well-recognised phenomenon[1] (Clancy *et al.*, 2011). Liquid fuel stoves are more prone to accidents than solid fuels and poor people tend to live in flammable structures, making the fear of fire prevalent. In Addis Ababa, Ethiopia, an ethanol stove is being tested as a substitute for kerosene. The stove has been well received by users and the prices of ethanol and kerosene are comparable (at least in urban areas). Ethanol as a fuel has performed well in terms of indoor air pollution and cleanliness (Murren and Debebe, 2006). Ethanol fuel in the form of a gel has been marketed as a safer option to the liquid. The gel fuel performed well under laboratory conditions. However, in South Africa, the commercial fuel was found to contain levels of water that could reduce the combustion performance. There was a lack of standards on water content, which was allowing poor quality fuels on to the market (Lloyd and Visagie, 2007).

However, in many rural areas ethanol as a cooking fuel would have to compete against fuelwood obtained at zero financial cost. Most rural households collect biomass. The poorer the household, the greater the reliance on biomass, which is collected from a variety of sources: forests, homestead, hedgerows, etc., and a variety of types are used, such as wood, agricultural residues and dung. The time invested in collection can be significant (estimates range from 2–20 hours a week) depending on availability, as well as the heavy loads carried. Poorer households tend to spend more time searching for biomass than higher income households (Clancy, 2002). Household energy provision is primarily a task for women, sometimes with the assistance of children, particularly girls. As fuelwood becomes scarce, or during busy times in the agricultural cycle, households have to take decisions about how to balance their assets and what coping strategies to adopt, such as collecting less biomass, using biomass found closer to the house (which may be of poorer quality[2] than from more distant sources), reducing the frequency of cooking and reducing the quantities and types of food cooked. Poor rural households buying energy carriers is not a common strategy (Cooke *et al.*, 2008). An explanation offered for the lack of transition to better quality energy carriers or improved cook stoves is that so long as the opportunity cost of women's unpaid labour remains low, they and their families will not consider time and labour saving a priority (Nathan and Nathan, 1997). Another financial barrier in the transition from wood fuels to biofuels for cooking is the need for the household to buy a new stove and possibly cooking pots to fit the stove. There are also issues of status. In some communities wealthy households demonstrate their economic status by the woman of the household not cooking – instead this is done by someone else and so there is no incentive to switch to improved stoves. On the other hand women who have professional jobs have a preference for their clothes not to smell of smoke (Sinha, 2012).

The issue of the energy carrier for food preparation is more than a financial one; the taste imparted to food when cooked with biomass for example is regarded as a major barrier to the transition to smokeless stoves (Clancy *et al.*, 2011). Food taste is a factor that ethanol will have to overcome.

On the positive side, ethanol can be bought in small quantities, unlike LPG, so fuel purchase can be matched to household cash flow. Ethanol also uses the same sort of distribution system as kerosene so there is no need to develop a complete new distribution system, helping to keep costs down.

The transition between energy carriers for lighting might be easier than for cooking because even poor rural households buy small quantities of kerosene or candles on a daily or weekly basis for lighting. However, there are fire hazards associated with kerosene and candles that can be a factor in making the transition to a safer energy carrier. Also it is considered that there are likely to be health effects from exposure to the particulate emissions from simple wick kerosene lamps. These effects are less widely recognised than those for particulate emissions from biomass cook stoves (Schare and Smith, 1995; Apple *et al.*, 2010). If the transition to a cleaner lighting fuel is to take place, a low cost ethanol lamp needs to be readily available. Such a lamp has been developed in India with a reported luminosity equivalent to a 100W electric light bulb (Rajvanshi, 2006). However, electric light is generally the preferred option for lighting, for reasons of safety, quality of illumination and simplicity of use (Clancy *et al.*, 2011).

Both bioethanol and biodiesel can be used to generate electricity (see next section). The benefits that electricity brings to rural areas are well documented (see for example, World Bank, 1996; Ramani and Heijndermans, 2003; Barnes and Foley, 2004; World Bank, 2008c). Villagers consider that electricity brings a feeling of modernity and opportunities for income diversification. The evidence shows that electricity is primarily used for lighting and TV – 80 per cent of consumption according to one estimate (World Bank, 2008c). It is rarely used for cooking – less than 1 per cent of the rural population (World Bank, 2008c) usually in areas where electricity is generated from micro-hydro (e.g. Nepal) and in East Asia, where electric rice cookers are popular. Refrigerator ownership is low – around 20 per cent of electrified homes. There are other benefits recorded, for example, the security that street lighting brings and the opportunity for economic and social activities such as music and dancing. Although much is made of the 'educational benefits' that TV brings, there are other benefits such as families sharing, for the first time, leisure time together (Massé and Samaranayake, 2003). The most significant benefits ascribed to electricity by rural people are that it makes home life more convenient and housework easier (Ramani and Heijndermans, 2003).

Improvements in household well-being are brought about by the use of electricity in health centres, for refrigeration and good quality lighting (particularly appreciated by women when giving birth). There is also some evidence suggesting that access to television and radio brings information into the household about nutrition and family planning.[3] Electricity (and engines run on biofuels) can run water pumps for both household drinking supplies and irrigation. The former would reduce women's time poverty and drudgery as well as reducing long-term skeletal damage linked to carrying heavy headloads (Matinga, 2010). Irrigation can improve agricultural output by increased crop yields, which could improve food security directly and indirectly through increased income from selling the surplus (see Chapter 5).

Lighting can also increase working hours for diversification in productive activities. Understandably there is concern about increasing the length of the working day, particularly for women. However, the evidence is mixed about the consequences of the longer day as a result of electric light or whether this longer productive day exits. A study of the effects of electric light in rural communities in Bolivia found that even after electrifying their households, people still followed the same daily routines based on the sun rising and setting, going to bed when it got dark (Sologuren, 2006). Some women appreciate the flexibility electric light gives in enabling them to organise their daily chores (HDRC, 2002) which can be considered to reduce the stress of having to do everything in daylight hours. Women are able to combine income generation activities with other tasks, such as selling cold drinks or making snacks to sell to people who come to watch TV (World Bank, 2008c). On the other hand, women and men enjoy the opportunity for spending time with their families, enjoying relaxation and entertainment such as watching TV and videos as well as listening to the radio (Clancy *et al.*, 2011).

It is recognised that the availability of electricity alone will not necessarily increase incomes or stimulate the establishment of microenterprises. Rural transformation using modern energy carriers needs support since entrepreneurs are uncertain about how to develop market opportunities and the role modern energy can play and the prospects this opens up (Kooijman-van Dijk and Clancy, 2010). By the early 1980s in Nepal, 28 district headquarters had been electrified by stand-alone systems. Ten years later, it was found that none of these supplies were being used for productive purposes, except for one photocopier (Aitken *et al.*, 1991). The need for complementary infrastructure, such as roads, transport, markets and banks, are considered important factors but other factors, e.g. gender and education level, also play a role (World Bank, 2008c). Telecommunications can open up new markets or allow price monitoring, thus enabling rural people to assess whether or not a trip to a market is financially worthwhile. Location close to an exploitable resource (either natural, such as good quality agricultural land, or of human construction, such as tourism) may also be significant. Population density also plays a role in providing a sufficiently large customer service base to allow a market for goods and services to develop (Kooijman-van Dijk and Clancy, 2010). While electricity allows for increased output and improved quality of goods, often its use is limited to improving the working environment of the enterprise through the use of fans, lights and TV or radio (Kooijman-van Dijk, 2008). It is possible that such enterprises may improve productivity while shops and eating establishments may attract new customers, but there is no guarantee of increased income.

Much of the emphasis from multi- and bilateral development agencies goes on electricity and there is less attention paid to the role of process heat in small enterprises.[4] There are enterprises linked to agriculture, e.g. leather working, food processing and textiles; while others are not, e.g. services such as shops and bars, and brick making. In general, these enterprises will all use process heat and it is quite common for rural enterprises to buy their fuel. Here biofuels could potentially compete with biomass more easily than for household cooking; provided the equipment is available, the heat output is sufficient for the process and any combustion products do not have an adverse effect on the product.

On the negative side, the production of biofuels has had an impact on cooking fuels in Indonesia where palm oil provides the staple cooking oil as well as the main feedstock for biodiesel. In 2007 the price the consumer paid for cooking oil increased by 40 per cent while the overall rate of inflation was 6.6 per cent (Bailey, 2008). The government responded to demonstrations against the price increases by providing cooking oil and soybeans to the poorest families at a cost of Rp500bn (US$54 million) (Bailey, 2008).

Community development

Decentralised energy supplies

Locally refined biofuels could substitute for petroleum fuels or be used to generate electricity in isolated rural communities. Such options are usually advocated on the grounds that locally produced biofuels can offer a lower cost alternative to grid extension and the transport of petroleum fuels. When serving local liquid fuel markets the biofuels must provide an energy carrier that not only offers advantages in terms of price and convenience but also matches the quality of petroleum fuels. Lack of adherence to fuel quality standards can damage the market. There are reports of palm oil being adulterated in Kigoma, Tanzania (Practical Action Consulting, 2009).[5]

Biofuels can be grown on a scale compatible with the level of demand likely in rural areas. In Mali, jatropha is being cultivated to supply the needs of surrounding villages. The jatropha is grown on communal land that villagers have collectively agreed to use for that purpose. There are also rental agreements with farmers for use of their land which can provide a source of income. The villagers see the benefits as a cheaper fuel which they are able to obtain much closer to home. Before the production of biodiesel, obtaining diesel required a 50 to 60 km trip to town and cost about 50 per cent of household expenditure (Cotula *et al.*, 2008). A cooperative in Mato Grosso, Brazil, grows biodiesel for its members' use and estimates savings of 40 per cent on pump prices (Bailey, 2008). However, a project in rural Sri Lanka to operate a Community Biodiesel Processing Centre to process 5 litres of biodiesel per day (which was considered sufficient to meet the daily basic energy and transport needs of the community) estimated the cost of one litre of biodiesel at around Rs300 (approximately US$2.25) while petroleum diesel was around Rs110 (approximately US$0.80) per litre. The cost of the processing chemicals (methanol, sulphuric acid and sodium hydroxide) was found to be significant in the total cost of production. However, this cost could be offset by selling some of the by-products (Musafer, 2010).

Biofuels can be used to generate electricity. Biodiesel can be used in a diesel genset and bioethanol can be used in a petrol genset. It is more common to use diesel engines than petrol engines for electricity generation. The former are more robust (with a lifetime of around 30,000 running hours compared to 5,000 hours for a petrol engine – these lifetimes will be shorter without regular maintenance) and are commercially available in a range of sizes starting at around 5kW. Petrol engines are more portable, since they are lighter than a comparable size diesel, and for electricity generation are found in the range of 500W to several kilowatts (Practical Action, 2006). Some typical rural applications for electricity and their power requirements are shown in Table 6.1.

A key issue then becomes: who is to provide the refined fuels or generate and supply the electricity? The community itself or the biofuel producer could become an Energy Service Company (ESCO). For community-run schemes the first challenge would be whether poor people would have access to information, advice and credit to enable the community to purchase appropriate equipment, the technical skills to install, maintain and operate the energy system in addition to the capacity to adopt standards and carry out quality control. Many small-scale energy systems have failed because the local technical expertise was not available (Mulugetta *et al.*, 2005). A

Table 6.1 End-use applications using electricity and typical power requirement

Application	Typical power requirement
Small-scale irrigation pumps	2–15kW
Small-scale electricity generation	2–50kW
Battery charging	500W
Grain milling or threshing	5–15kW

Source: Practical Action, 2006

second factor which has rendered many community schemes unworkable is revenue collection. There have been considerable difficulties in dealing with defaulters or the unit charge has been set at a level that results in the revenue collected only at best covering operating costs, with no capital reserve to pay for maintenance and replacement costs. In Mali, the state-owned rural electrification company is involved in a project that uses jatropha oil grown by small-scale farmers to generate electricity for powering a 300kW micro-grid. The responsibility for generation and electricity sales lies with a private company, ACCESS. At the end of 2008, 247 houses were connected to this grid and street lighting has also been installed. There is a connection fee of US$30 and a differential tariff system for low and high levels of consumption. ACCESS has managed to collect over 90 per cent of bill payments and covers nearly 100 per cent of operating costs but at the same time has taken a sympathetic approach to those having problems paying bills[6] (Practical Action Consulting, 2009).

There are also other difficult 'management' issues to deal with, which communities may not be well equipped for, such as resolving disputes relating to dissatisfaction with or misuse of the service. Small decentralised grids have a number of technical problems, for example, the voltage can vary significantly when equipment switches on and off and fluctuating voltage can cause malfunction of or damage to electrical equipment. In addition, the capacity of a decentralised grid is limited and unlike the national grid cannot absorb the use of large numbers of equipment which draw a lot of power, e.g. refrigerators. Such a system requires the cooperation of the users to abide by the rules of what can and cannot be connected to the grid.

There has been a considerable amount of research in Brazil on developing micro- and minidistilleries.[7] It is considered possible to integrate small-scale distilleries into a smallholder farming system with the potential to make the farm more self-efficient. As well as producing a fuel from sugarcane, the bagasse and molasses can be used to feed cattle and the effluent from distillation can be used as a fertiliser (Ortega *et al.*, undated). Ethanol standards for household-end uses, such as cooking and lighting, do not need to be as high as for blending with petrol. Unfortunately at the time of writing, the law in Brazil does not allow for the direct sale of fuel ethanol by individual farmers from micro- and minidistilleries, although it can be marketed through a cooperative (Couto and Murren, 2009).

Both bioethanol and biodiesel are new energy carriers to most consumers even if the biofuels have very similar properties to their fossil fuel counterparts. This requires market development by the ESCO supplying the energy carrier. Communities operating an ESCO will probably have to enter uncharted waters, which may require outside help. In the state of Minas Geras in Brazil, an ESCO has been involved in a microdistillery project, which has multiple benefits for smallholder households who are supported to grow sugarcane for ethanol and other by-products for sale. The ethanol is intended for use as a clean cooking fuel. Initially the ESCO has supplied the stoves to 39 families, and these families are entitled to a monthly quota of ethanol at a price

below the market rate. The long-term aim is to gradually expand production and support communities with the development of the microdistilleries
(Practical Action Consulting, 2009). Whether such schemes can be economic
and competitive with fossil fuels, even for use in remote locations, depends
on the relative prices of agricultural feedstocks and petroleum fuels, as well
as technological developments (FAO, 2008a). Potential threats to fuel supply
exist if there are competing buyers from other value chains. For example, the
tax breaks on imports of spirit quality ethanol offered by the EU provided
a more lucrative market than the domestic market for Zimbabwe's ethanol
distillery. This switching markets is considered to have played a significant
role in the demise of bioethanol/petrol blends in Zimbabwe (Kartha *et al.*,
2005). Biofuel crops are also vulnerable to unpredictable events such as the
weather and outbreaks of pests and diseases, which can lead to crop failure,
and without sufficient storage of fuel reserves, energy shortages can occur.
The Zimbabwe ethanol programme was severely affected by droughts in the
early 1990s (Kartha *et al.*, 2005).

Large-scale agro-processers, such as sugar refiners, do supply electricity to
residences within their estate boundaries. However, the law usually forbids
third parties from generating and distributing electricity. So for large-scale
agro-processers and biofuel refiners to supply the community requires regulatory reform in the energy sector, which is a process in a broader context.
The regulatory framework that exists in many countries is often considered
to create a major barrier to the development of decentralised energy facilities by anyone other than state-owned utilities, who in turn are reluctant to
undertake such ventures due to the high maintenance costs. Existing regulatory frameworks have been described as 'hostile, contradictory, or uncertain'
(IDS, 2003: 35) which stifles private sector finance and innovation.

Public–private partnerships for community services

There is a view that companies should, as a matter of good practice, integrate
social and environmental concerns in their business operations and in their
interactions with the people in whose communities the companies operate.
This approach can go beyond good employment and environmental practice
and include the provision of local community services as a contribution to
addressing rural poverty. This view is encompassed in what is commonly
known as corporate social responsibility (CSR).[8] On the other hand, there are
those, primarily from the business community, who consider that it is not the
role of business to contribute to poverty alleviation other than by providing
employment and indirectly by paying taxes. The building of schools and
clinics and infrastructure is the role of the state (Newell and Frynas, 2007).

Companies can see that building infrastructure, such as a road and tele-
communication systems, will assist their business directly while building
local health clinics and providing clean water can mean healthier employees.
These are facilities that all members of the community can potentially benefit
from, although they may not have automatic rights of use. However, the

long-term sustainability of such ventures can be problematic with communities being left to finance the day-to-day running costs, which can prove to be beyond their financial and managerial capacity.

Most of the empirical evidence related to CSR appears to be related to conditions of employment and environmental protection. There are examples of external investors in biofuel production going beyond this, for example, an IFAD funded project in Uganda is a partnership between the private sector and small-scale farmers (see Box 3.2) which promotes health care for its employees by providing healthy meals and a clinic. Developers do claim to provide such social goods under sustainability objectives and indicators, e.g. under CDM, however, independent verification of such claims is difficult to find (Lee and Lazarus, 2011). As was pointed out above, biofuel refiners are often able to supply electricity in areas where it is too expensive for the grid to reach although existing legislation can act as a barrier to third parties selling electricity to consumers.

Reduction of drudgery and time poverty

In Chapter 2 poverty was shown to be more than a lack of income. Poor people lack many things which could help them move out of poverty, and these deficiencies are often interlinked. Time was one asset that poor people were particularly short of since fulfilling many of their basic survival tasks are based on their own labour. These tasks are often arduous and involve drudgery. Time poverty is closely linked to energy poverty. Solving energy poverty by providing sufficient quantities of good quality, affordable energy carriers can make a significant contribution to ending time poverty and reducing drudgery. Biofuel production in rural areas can play a multiple role in addressing energy poverty. First there is the opportunity to improve incomes which would allow for the purchase of better quality fuels. However, it is important in this regard to make sure that there is gender equity in the access to income-generating opportunities from biofuels because men and women have different energy priorities at the household level. It is not unknown for a rural household to get an electricity connection which the men use for watching TV while the women still have to spend long hours collecting fuelwood (see for example, Sparknet, 2002; Clancy *et al.*, 2011).

A study in Sri Lanka found that women's time saving is the major benefit of a household electricity connection. Eighty per cent of the interviewees reported saving between one and two hours through avoided journeys (such as taking batteries to be recharged, and going to the city to buy kerosene, medication and vaccinations) and on household activities (such as firewood collection, cooking, ironing, boiling water, house cleaning and chimney cleaning) (Massé and Samaranayake, 2003). It is interesting to note how women made use of their extra 'free time'. Of the female household members, 29 per cent said that the time they saved was spent on extra housework, while less than 5 per cent reported using it for productive activities.

There is also the prospect of decentralised energy production using biofuels which improves energy supply in rural areas. There are a number of projects with that objective that are also linked to drudgery reduction. One of the most well known is the Multipurpose Platform Project which first started in Mali and has now spread to other African countries. The project is a decentralised energy system based around biofuels. It was established in Mali at the request of rural women's associations to find appropriate and affordable substitutes for their own energy, so that they can engage in activities that generate income, and that provides benefits for themselves and their families (Burn and Coche, 2001). The platform consists of a small diesel engine, running on biodiesel produced from locally grown jatropha, which is mounted on a chassis. A variety of end-use equipment can be attached, including grinding mills, battery chargers, vegetable or nut presses and welding machines. It can also support a mini grid for lighting and electric pumps for a small water distribution network or irrigation system. The mechanisation of grain mills has been found to be one of the major gains for women in terms of time saving and reduction of drudgery in preparing staple foods (Clancy and Kooijman, 2006). In Ghana, a shea butter processing project which uses mechanised oil extraction equipment based on a biodiesel fuelled diesel engine[9] has resulted in considerable improvements in output (from 25kg butter in two weeks to 1 tonne in four weeks) and a reduction in drudgery (Karlsson and Banda, 2009).

There is a distinct gender dimension related to time poverty. The influence this has on participation in biofuel projects is discussed later in this chapter. However, if we look at biofuels as energy carriers, there is the potential to make a significant contribution to reducing women's time poverty in particular. As stated, women spend many hours collecting fuelwood or going to town to collect kerosene and batteries. If biofuels can substitute for fuelwood or provide affordable electricity then they can make a significant contribution to rural development.

Food security of the rural poor

While it might not be surprising to learn that one-fifth of people suffering from food insecurity are landless, since they do not have one of what are considered essential assets for food security – ownership of land (see Chapter 5), it is probably surprising to learn that half of the population suffering from food insecurity are smallholders, people who have access to land. We do not intend to discuss here the reasons behind these findings, but instead we will examine whether or not biofuels increase or decrease the food security of the rural poor, and hence move people out of or into (deeper) poverty. Biofuel production can impact on the three mechanisms by which the rural poor obtain their food:

1   growing their own
2   buying supplies in the market
3   harvesting ecosystem services.

In terms of growing their own food, smallholders need to retain owner-ship of land and they need to maintain or enhance the fertility of that land. There is evidence to show that biofuels have been changing patterns of land ownership, particularly in Latin America and in parts of Asia, in particular Indonesia and the Philippines. There is an expanding literature on what has become known as 'land grabbing', where agro-businesses acquire large swathes of land for crop production; although we would stress again that this is usually for food crops, but we recognise also that it could easily change. To what extent smallholders sell or relinquish their land voluntarily is contested. We have cited in Chapter 5 examples where small-scale farmers have been violently deprived of their land, particularly in South America and Indonesia, and in other instances where some dubious land rental deals have been nego-tiated that far from benefit the owner. In Africa, large-scale land acquisition has tended not to displace smallholders because the land allocated for biofuels development is classified as unproductive. However, the use of this land has other repercussions for food security which we discuss in the section below. On the other hand, it would be wrong to underestimate the lure of the city for what appears to offer an easier way of making a living than the drudgery of dawn to dusk under the hot sun in the fields. Novo and his colleagues suggest complex reasons, other than economic calculations of costs and benefits, why farmers in Brazil decide to rent out their land to large-scale producers (Novo *et al.*, 2010). For smallholders, keeping the link with the land is culturally important to maintain their identity as farmers. Leasing instead of selling retains the option to return to farming at a later stage, while a long-term contract with a large company offers a type of stability that individual efforts at participating in the market does not. In addition, farmers who lease land can also remain members of the sugarcane farmers' association and still have access to benefits such as credit and health care, which they would be denied if they sold their land.

Land-use changes influence food availability, both in terms of types and quantities of food available. The available evidence shows that large-scale plantation biofuel crops have had the greatest impact on rural people's food insecurity. Plantations have displaced smallholders from their land and biofuels have displaced some commercial food crops. This affects not only rural people's capacity to feed themselves but also local food markets in terms of availability of staples and prices. Large-scale plantations of agribusiness biofuels can also place a high demand on water use, which in turn can have impacts on local water availability for other users. However, there is little empirical evidence to show whether or not there is an actual impact on food production

Incidents have been reported of small-scale farmers who retain ownership of their land switching from subsistence crops or crops intended for the local market to biofuels (e.g. in Tamil Nadu (Ariza-Montobbio *et al.*, 2010) and in Swaziland (Sheil *et al.*, 2009)). This switch has had an impact not only on food availability but, depending on the crop, also on fodder or firewood availa-bility. It is rather surprising that farmers completely switched to an unfamiliar monoculture crop because they tend to be risk averse. However, most of the displacement is linked to jatropha where the farmers seem to have been given unrealistic expectations of yields and hence income. Once reality sets in, disillusioned and sometimes poorer farmers have uprooted the plants and returned to their traditional crops.

A number of projects involving smallholders report the potential for enhancement of food availability where attention has been given to providing training in crop management, including intercropping of biofuel crops with legumes and the use of organic fertilisers, with the aim to improve yields. However, such statements await quantification and a long-term assessment of outcomes.

If biofuels are used to fuel rural energy systems, they can contribute to improving food availability by helping reduce postharvest losses through improved processing technologies (as demonstrated in Box 5.3 relating to the shea butter project in Ghana) and long-term storage. Access to such technologies would increase the volume of food available and help to bridge the hungry periods when a household's own food production is low. In addition, if biofuels provide an affordable rural energy supply, the well-being of poor households can benefit from access to clean drinking water (through pumping and sterilising) and from cooking food as well as the prospect of food storage through refrigeration.

The majority of rural households buy some food and up to half of all household expenditure can be on food. Biofuel production can offer a chance to increase household income. However, as Amartya Sen pointed out, who controls the disposal of income is a key factor in household food security (Sen, 2001). The evidence suggests that the greater the degree of control women have over the family income, the greater the proportion of income spent on food. Therefore, one way that biofuels can contribute positively to household food security is by enabling women to participate in production chains, although participation is not enough to guarantee food security. Much depends upon the terms of incorporation as well as intra-household decision making and power relations within the household (see section on gender and biofuels).

It would appear that seasonal plantation workers are particularly likely to be food insecure because their wages can be low relative to other estate workers. They often do not qualify either for small plots of land or vegetables supplied by the estate owners, and food bought on the open market can be expensive. The other way that wage labour has been affected by biofuels is when biofuel crops have displaced crops that are more labour intensive, e.g. in Colombia palm oil replaced bananas; or where there has been increased mechanisation, e.g. in the sugarcane fields of Brazil.

Biofuels seem to be having the greatest impact on undermining food security where land classified by government as 'waste' or 'non-productive' is allocated for large-scale biofuel production. There seems to have been a complete lack of understanding by urban elites and biofuel developers about the role of this land in rural people's livelihood strategies. Not only does this land play a direct role in household food security by providing food and fodder, as well as other necessities such as medicines which support well-being, but it also offers an indirect role through income generation opportunities for household members by utilisation of the resources. It is the most vulnerable of rural households, the landless, that place great reliance on these ecosystem services. It is also some of the most food insecure areas (Africa and India) where 'waste' or 'non-productive' land is promoted for growing biofuel crops. The relation between food security and ecosystems is discussed in more detail in the next section.

Ecosystem protection

There are multiple drivers of ecosystem change, and biofuels is only one of these. However, there is no doubt that biofuel production from first generation feedstock has a bad reputation with many people due to the association with impacts on biodiversity from land clearance, particularly tropical rainforests. As has been shown in Chapter 4, the situation in relation to biodiversity and biofuels is complex and it can be simplistic to blame biofuels for environmental degradation at a specific location, including threats to habitats and their associated diversity of ecosystems. The level of threat will depend on the local conditions, in particular the characteristics of the existing land cover (natural forest, managed forest, crop land and fallow land), where biofuels are grown. This book focuses on the rural poor and has focused, therefore, on the direct and traceable impacts biofuels can have on their lives, so we have not entered into the debate about whether or not biofuels enhance or mitigate climate change.

There is consensus that the vulnerability of the rural poor can increase if the ecosystems they rely on to provide goods and services are diminished. However, there is very little empirical evidence about the way that ecosystems respond to change and the way this influences the level of ecosystem services. There is also limited understanding about the extent that the rural poor use these services. Also such information needs to be ecosystem specific and disaggregated to reflect the social characteristics of a community, particularly for gender because it appears that women draw more than men on these services. Such information is the basis of sound policy making and pro-poor interventions. The Millennium Ecosystems Assessment project is a step in the right direction in providing data for more informed decisions. However, even without that data, it is still possible to make a general statement that decisions about growing biofuels should not result in the situation where the rural poor are excluded from access to ecosystems they have customarily used. There are lessons to be learned from some wildlife conservation projects, where well-meaning powerful outsiders set the agenda and have large areas of land enclosed, with local people denied access to traditional resources and cultural sites (Swiderska *et al.*, 2008). There is increasing recognition of the need for a more inclusive and ethical approach in natural resource management, which recognises the devastation such conservation approaches have on rural communities as well as the constructive role rural people already play in managing the resources on which their livelihoods depend.

Biodiversity is the basis of many ecosystem services both from natural habitats and agricultural systems. Indeed there is a strong linkage between the biodiversity in natural habitats and agricultural production, for example pollinating insects may forage among crops but shelter in natural vegetation. The main threats to biodiversity and ecosystems from biofuels, which can result in negative outcomes for the rural poor, come from land-use changes that reduce biodiversity and affect water quality and use. It is possible to categorise the land that could be converted to biofuel production into three types: land with natural vegetation, such as forests and savannah; non-productive land, that is land which might or might not have been used for crops; and land currently used for crops. The types have different levels of biodiversity and provide different types of ecosystem services but all play a part in the lives of the poor. The challenge is to ensure that all three types maintain their ecosystem services.

The international responsibility for protecting biodiversity is set out in the United Nations Convention on Biological Diversity (UNCBD, 1973). The Conference of the Parties to the Convention has recently begun to explore possible options for addressing the potential impacts of biofuels on biodiversity, which are in essence seen as similar to those of modern agricultural systems. A recent report prepared for the Conference on biofuels and diversity concluded that biofuels could be integrated into their work on forest biodiversity and on agricultural biodiversity (UNEP, 2008). However, there is a significant gap between what is promised at the international level and what is delivered at the local level. The Ecosystems Approach[10] was designed to protect ecosystems and their services through integrated management of land, water and living things. The approach takes into account social and cultural issues by emphasising multiple needs, decentralisation, community rights and cross-sectoral integration. It appears that it is mostly absent from national plans to preserve biodiversity. Biodiversity tends to be treated in its totality rather than focusing on the components of biodiversity that are most important for the livelihoods of the rural poor. Further cause for concern includes the attention in practice being given to protect areas rather than areas of economic activities, such as agriculture (Swiderska *et al.*, 2008). This would imply that land appearing unproductive to outsiders would be particularly vulnerable to exploitation, since the value of biodiversity is neither recognised in general, nor the specific role of such areas in rural livelihoods by those who make the decisions to sanction biofuels development. There are institutional weaknesses including: decision makers lacking information about the links between poverty, biodiversity and ecosystem services; fragmented coordination among many different agencies, which leads to absent or weak accountability mechanisms for the state; and business decisions about ecosystem services (Irwin and Ranganathan, 2007, cited in Swiderska *et al.*, 2008: 42). There is also a lack of rights and voice in the decision-making processes for rural people who are best positioned to explain the role of ecosystem services in their lives, although indigenous knowledge about ecosystems is generally ignored by outsiders (Millennium Ecosystem Assessment, 2003; Swiderska *et al.*, 2008).

Biofuels compete with drinking water as well as water for crops and, probably less well recognised, for natural vegetation. Agricultural already takes 70 per cent of the fresh water withdrawn from rivers and groundwater, and increased biofuel production would probably increase the demand (Global Bioenergy Partnership, 2011). The overall effect of this competition could be reduced yields of crops and a decline in natural vegetation which form important sources of food and medicines for people and animals.

In respect of water quality, biofuel production can lead to pollution due to run-off from excessive fertiliser and pesticide use, although this can be a feature of any intensive agricultural system. Discharge from processing plants can also be problematic. In both cases drinking water quality is affected as well as aquatic life, which can affect protein sources and the livelihoods of rural households. The residue from palm oil processing can be highly polluting when it is dumped in waterways. However, technologies are available to treat the residues, often producing valuable by-products which offset the investment costs. For example, a project using palm oil residues has been registered under the CDM for a number of sustainable development benefits, including improved water quality and increased energy security (Lee and Lazarus, 2011).

Ecological damage occurs not only as a result of land-use changes due to large-scale plantations from biofuels. Under certain circumstances, small-scale farmers are not always able to farm in a sustainable way, for example, when farming marginal land, they cannot always afford the inputs to improve the soil quality (Reardon and Vosti, 1995). Yet, by growing biofuels, they can also contribute to improving agricultural ecosystems. By inclusion in biofuel projects, smallholders are gaining access to techniques for improving soil quality such as enhanced fertility through increased organic matter, soil erosion and drainage/water retention. Farmers could use such knowledge to help improve food output. Evidence in Chapter 5 has shown that smallholders are keen to participate in projects for growing biofuel crops when it allows them access to agricultural information on cultivation techniques, as well as the chance to diversify income sources. However, such knowledge transfer requires a committed project developer with the aim to upgrade skills along the value chain. The existence of such schemes is a function of the value chain and the motivation of chain actors (see Chapter 7). The standards developed for more ecologically friendly biofuel production are aimed at large agribusinesses and there is no requirement for small-scale producers to meet such standards. On the one hand, the cost of meeting such standards can be prohibitive for small-scale farmers so there could be resistance to compliance. On the other hand, this non-inclusion in compliance with standards opens up an opportunity for developers to avoid compliance by outsourcing to smallholders.

Gender and biofuels

The concept of gender provides an analytical tool to assess whether or not women and men are affected equally or equitably by a process, such as the focus of this book: biofuel production and utilisation. What do we expect for women and men in terms of participation in biofuel production chains? What do we expect for women and men in terms of biofuel utilisation? Pro-poor approaches should be gender sensitive, recognising the generally unequal position of women and men in terms of ownership and access to assets. Certainly we are looking for signs that households are at best moving out of poverty, and at worst are not moving (deeper) into poverty as a consequence of biofuels. So in terms of gender we would be looking to see how the relative position of women-headed to men-headed households has changed as a consequence of biofuel production. Gender equity would require that women should have the opportunity to participate in biofuel production for which the terms of inclusion should be the same as for men. In terms of intra-household impacts from biofuels, women should not be disadvantaged by the loss of access to and control over household resources, e.g. land and cash, which are diverted to biofuel crops, thus preventing women from fulfilling their household responsibilities. A gender analysis would also look for indications that women have been empowered through participation in biofuel value chains in the sense that there is a transformation in gender relations and women gain greater control over their lives.

Participation in biofuel production creates opportunities to earn income and gives access to other benefits. However, based on experiences with agricultural production in general, it is unlikely that women would benefit equally with men in terms of participation, since the level of women's key assets for participation, ownership and control over land – which provide access to credit, as well as literacy and technical skills – are generally less than men's from the same community (see Chapter 2). Of the empirical evidence that is available, there are gender differences in willingness to participate in biofuel projects and hence how benefits accrue. The distribution of intra-household income is important if women are responsible for household food security. Men's motivation for participation is influenced by the levels of income to be earned and opportunities for access to new knowledge and skills as well as credit. These opportunities would ordinarily be difficult for smallholders, but for women, who have more restricted options for income generation compared to men, participation in biofuel value chains presents an opportunity to earn cash income close to home that makes inclusion, even on adverse terms, attractive and they are prepared to occupy the space. Women also value other benefits, such as access to the by-products from the crops.

Women's lack of skills, time poverty and lack of ownership and control over land tend to confine them to inclusion at the bottom end of the value chain, in low quality more poorly paid labour market jobs. There are examples, however, where project developers have consciously targeted improving women's skills, providing them with new knowledge, enabling them to have better paid jobs, which has transformed their status within their families and communities (see for example Karlsson and Banda, 2009). On the other hand, the numbers of these low skilled jobs on plantations are decreasing through mechanisation. The lack of sex-disaggregated data prevents us from tracing the dynamics of employment to determine whether or not men are being displaced by women in low skilled jobs.

Another area in which there is a lack of data relates to women's time poverty and biofuel production, both in terms of whether or not they have sufficient time to participate, and if they do, the effects of participation on their overall workload. There is no empirical evidence from biofuels projects to show the effects of time poverty on women's level of participation. However, we can draw on experience from food security projects, which biofuels might be expected to mimic, in South Asia where women are given the opportunity for on- and off-farm income generation. In such projects women's capacity to participate is severely constrained by their already long working day (16 hours in the study areas of India and Nepal) and so participation requires time management, adjusting the household chores to fit with income generation activities (Rahman Osmani, 2010). More worryingly, there was an increase in women's time burden relative to men's in households where women's control over household assets had increased. Husbands either withdrew partially from income-generating activities or refrained from increasing their involvement. Indeed, men in this group appeared to enjoy two hours' extra leisure compared to men whose wives had not gained control over assets.

As was shown in Chapter 3, while women are more readily prepared to participate on adverse terms than men, they do not passively accept the terms of incorporation (South Africa) or the impacts that biofuel programmes can have on ecosystem services (Ghana) and their social and political status (Brazil). There are indications of transformation in women's lives where the access to income is changing their status and providing them with a voice in value chain governance (India). Even so, this voice is limited to the bottom end of the chain and this space is not necessarily created voluntarily by men but as a consequence of a legal requirement of reserved seats.

Of the empirical evidence that is available, it would seem that women are disadvantaged by the establishment of large-scale monoculture biofuel plantations. These plantations can cover thousands of hectares with monocultures eradicating natural ecosystems (see Chapter 4 and above), which form an important source of goods and services for rural households, particularly for women (Kartha and Larson, 2006). Women are also vulnerable to displacement from land they farm due to gender inequalities in ownership and control over land (Cotula *et al.*, 2008).

In terms of utilisation of biofuels as an energy carrier, evidence shows that there are equitable benefits for women and men in the form of local energy services (as discussed above). Here, then, is potential to reduce the time poverty and work burden experienced by rural people, particularly women.

Income generation and its wider impacts

Biofuels do offer smallholders an opportunity to diversify their income. There are different arrangements for remuneration for participation in the value chain, some of which benefit small-scale farmers, and there are examples where they have not done so well. However, biofuels are a new crop not without some risk and smallholders are generally considered risk averse (Dauvergne and Neville, 2010). There has also been some price volatility in commodity prices which could expose those with little to fall back on to a relatively substantial risk, particularly where they have been persuaded to focus on a single commodity. There is evidence of this happening, for example, in Swaziland. However, there are examples of best practice where farmers are made aware of the risks of participation from the beginning. The Hassan Biofuels project makes it clear during the initial discussions that the biofuel crops are to supplement and not displace existing crops. They will not become rich from growing non-edible oilseeds but the aim is to identify sufficient indigenous species which fruit at different times of the year to give a steady income throughout the year. Expectations, therefore, are not raised to unachievable levels. At this point some farmers opt for exclusion (Narayanaswamy, 2009).

Fair contracts may provide the possibility for stable incomes. Farmers in Tanzania growing palm oil have shown an improved income since the increased interest on the world market in palm oil for biodiesel. Around the Kigoma region, the price paid at the local market[11] for a 20 litre container of crude palm oil in 2008 was US$15.30–20.40 which had risen from US$1.60 ten years earlier (Practical Action Consulting, 2009). The amount the farmer receives is slightly less since they have to pay local taxes and the middleman takes a percentage. Participants in the value chain benefit not only from increased income but there are also opportunities to build their knowledge and skills levels. Social networks also increase by participation in cooperatives or producer groups.

However, it takes more than fair contracts for smallholders to be able to participate in biofuel production. The nature of any processing equipment should not require high levels of technical education because there are still high levels of illiteracy, especially for women. Small-scale processing equipment is available and biofuels can be produced for local use. Supplying national or international supply chains requires a level of technical skills to meet quality standards, which would likely yield a better remuneration than growing the crops.

There are two broad ways in which biofuels can influence rural incomes other than direct involvement in growing and processing biofuel crops. First where the biofuel value chain allows for diversification in products and services at the local level, the opportunity opens up for other actors to participate and benefit. The study reported in Box 6.1 shows how it is possible for households to earn additional income from different by-products related to biofuel crops. Access to these by-products, or even the oil itself, is also a

---

**Box 6.1 Income generation by rural women and men
contributing to biofuel value chains**

The Biofuel Park programme in Hassan district, Karnataka, India, was established
to integrate the state's rural population into India's national biofuel programme.
The programme provides villagers, free of cost, genetically superior, mainly
indigenous, saplings from a range of non-edible oilseed-producing plants at the
onset of the monsoon season. The villagers can also rent a mechanical oil expeller
from the Biofuel Park or buy one from the state government at a subsidised rate
(Rs2,500 rather than Rs5,000).

The project activity discourages using agricultural land for biofuel crops and
encourages the use of bunds and hedges by men as well as household backyards by
women. Seeds are promoted as a supplementary income source. Such an approach
avoids interfering with food production. The bunds that run across fields and the
hedges surrounding fields retain residual moisture and are hence ideally suited for
rain-fed irrigation, although the saplings do need to be watered in the first year.

Narayanaswamy estimated the quantity of biodiesel likely to be produced
in Hassan district and hence estimated the potential income from growing the
oil seeds. In the tenth year after planting, the maximum yield is approximately
340,000 tonnes of oil which would sell for US$3.4 million. In 2001, there were
approximately 308,505 households in Hassan district. Therefore, the average
annual household income from the oil would be US$11. However, this could be
increased by selling the oil cake (approximately US$10), plant prunings (approxi-
mately US$15) and carbon credits (approximately US$1). While these figures can
only be taken as indicative (since they depend on assumed selling prices), the
annual total income per household (US$37) is similar to what can be earned from
an acre of good agricultural land.

Source: Narayanaswamy, 2009

---

motivation for participating in schemes growing the crops. As was pointed
out above, sale of the by-products can also positively affect the economics of
production.

The second indirect way in which biofuels influence rural incomes is
where biofuels increase local energy supplies, and the opportunities for
income generation also increase. Agricultural output can be increased
through irrigation (Nepal) and higher quality products through mechanised
milling (Ghana) (Karlsson and Banda, 2009). Villagers can receive training
to operate and maintain equipment, for example, the previously mentioned
Mali project. The Indian biofuels programme makes oil expellers available so
that producers can expel their own oil.

Another indirect income source available to the South involves trying
to attract investment in biofuel projects from the carbon finance markets
using the market value of expected GHG emission reductions. One option
is the Clean Development Mechanism (CDM), established under the Kyoto
Protocol, which allows industrialised countries to invest in projects that
reduce emissions in developing countries in return for certified emissions
reductions which are used to meet industrialised countries' Kyoto Protocol

emission targets. However, whether or not this is an option for small-scale farmers is doubtful. First, the transaction costs associated with registering a project are considered prohibitively expensive (Bakker, 2006). Once the project is registered there is then the need to monitor it, which can be expensive when the project involves many small-scale farmers. It is also a moot point as to whether or not CDM project developers would be interested in biofuel projects involving large numbers of small-scale farmers, since they are likely to be searching for the highest return per dollar invested and low management costs. An analysis of CDM projects, as of January 2010, either registered or in the validation stage was able to identify only one project linked to liquid biofuels (Lee and Lazarus, 2011).

An increase in rural incomes is considered to create a virtuous circle of development. As cash surpluses accumulate the possibilities for investment in agricultural production increase, which creates a demand for goods and services. These can generate local opportunities. However, it is thought most likely that poorer local households rather than higher income households will purchase their goods and services locally (Kartha *et al.*, 2005). This supports the argument that suggests that biofuels programmes that are pro-poor can make a significant contribution to rural development, although there are more inputs needed than growing and processing biofuel crops to achieve such a transformation. Indeed, particularly in the more remote areas, considerable support will be needed to stimulate entrepreneurism as well as ensuring that key infrastructure, such as roads and telecommunications, is available. Capacities need to be developed in business management, marketing skills and negotiation. An example of failure to take a holistic approach into account is the Ghana project. Mechanised shea butter processing, using biodiesel-fuelled presses, increased the volume of output with the knock-on effect of driving prices down, which although good for consumers was detrimental for the producers (Karlsson and Banda, 2009).

Biofuels and rural poverty

At the end of Chapter 2 a number of questions were posed about the effects of biofuels on rural poverty: to what extent do biofuel production chains really provide an opportunity for the rural poor? What form does participation take? Who is benefiting and who is losing out? How? In this section the evidence from Chapters 3, 4 and 5 is brought together to answer these questions. The lack of objective empirical evidence makes definitive answers to these questions difficult. As pointed out in Chapter 1, the available data is often provided by stakeholders with specific agendas, such as project developers or environmental NGOs, so it is difficult to verify some of the claims. In addition, there are limited long-term biofuel projects to draw on for data so some of the indirect benefits, such as the provision of health centres, might yet materialise. However, it is possible to give indications based on the available evidence about the pro-poor nature of biofuels, and in some cases draw on experiences from agricultural projects which biofuel crops mirror.

Biofuels are allowing a diversification of rural income sources both as an agricultural product and by supporting non-farm activities as an energy carrier. Rural people are able to participate in biofuel value chains in a variety of ways, from directly growing crops on their own land (contracted or open market sale), as wage labourers, renting their land to developers, and to a lesser extent, processing crops to fuels (or to an intermediary product). Rural people are very much confined to the low value, bottom end of the chain. However, we live in a more interconnected world and rural people are more aware of what is possible and are prepared to challenge for better terms of inclusion, for example, South African women's reaction to the government's biodiesel programme (Karlsson and Banda, 2009).

This book focuses on the rural poor who are identified as people whose land, labour or cash are at such a level that they are constrained to make investments in securing a sustainable livelihood (see Chapter 1). The poor are not a homogeneous group but exhibit a complex social identity and variation in ownership of assets, and possession of skills which can influence the possibilities for participation in biofuel value chains either as producers or labourers. These social factors are not of general interest to market-based initiatives, so it is not surprising that it is the wealthier rural households that are able to participate in the opportunities offered for income diversification. However, there are examples where social factors are taken into account to promote inclusion either by governments, such as India's biodiesel programme in which women and the landless are specific target groups for participation, or by NGOs and other non-state actors, such as ARC Kenya, an NGO that is stimulating the participation of young and blind people to grow jatropha (Amezaga *et al.*, 2010). ARC Kenya's initiative promotes inclusion of groups identified in Chapter 2 who are likely to be excluded: disabled people, who face particular barriers to participate in markets, and the young who are reluctant to engage in farming. However, it is not necessarily the case that private sector = bad biofuels, NGOs = good biofuels. In Indonesia, schemes to promote growing jatropha by smallholders were introduced by a local NGO and by a multinational company. In both cases there were overestimates of the likely income for the farmers. The multinational company compensated the farmers with free fertilisers and stoves while the NGO offered no compensation. In the former case the farmers continued to participate in the schemes whereas in the latter they opted for self-exclusion (Fatimah, 2011). Such compensation to smallholders is considered important for covering gaps in income owing to reduced harvests when new crops are planted, such as jatropha, which takes two or three years to mature,.

There is a lack of empirical data about the income levels farmers are earning from growing biofuel crops, therefore it is difficult to make pronouncements about the direct impact biofuels are having at least on the financial dimension of poverty. There are farmers who consider the levels of remuneration sufficient from growing biofuels, while others, even within the same scheme (e.g. Hassan Biofuel Park), consider that the levels of remuneration are not sufficient. The reasons for inclusion or exclusion of the poor in biofuel chains are complex. A common explanation is that people with limited assets are often risk averse and so they are reluctant to take decisions that might tie up labour, land or financial assets. This may be too simplistic an analysis (Hospes and Clancy, 2011). It is assumed that inclusion is wanted by the excluded but, as we have seen from women in Brazil, this is not necessarily the case. The women opt for exclusion because they consider that inclusion works against their social and political status. Furthermore, income is not the only reason to opt for inclusion: it is also found that smallholders' access to resources, such as loans, fertilisers and agricultural extension support are hard to obtain by other routes.

One of the most worrying aspects of biofuels is the effect on ecological systems, such as soil fertility, water and biodiversity. Rural people, particularly the poor, draw on these natural assets for a range of goods and services. The impact on food security both directly, as a source of household nutrition, and indirectly, as a source of income to buy food, is particularly problematic. Large-scale biofuel crops grown by agribusinesses are definitely competing with smallholders and other rural residents for natural assets, particularly land and water. The type of land and the mechanisms of acquisition vary as do the effects. In Latin America good quality agricultural land is being taken over often by highly contested means, and smallholders are losing their livelihoods. On the other hand there are smallholders who are leasing land on terms they consider acceptable. In Africa and India, governments are promoting the use of 'non-productive' land and leasing large areas to foreign investors. The term 'non-productive' (or its variants) is also highly contested. There seems to be a singular lack of understanding of the basis of rural livelihoods by outsiders. The use of 'non-productive land', frequently promoted so as not to undermine food security, seems to have had the reverse effect and deprived the poor of their safety net. At least some external developers have recognised their mistakes when pointed out by villagers (see Box 5.2) and withdrawn their projects.

Work on the water footprint of biofuels shows gaps in our knowledge of the water demand to produce economic yields of many of the plants being considered as feedstocks. There have been a lot of over-optimistic statements about the yields of plants about which there is little agronomic data under varied conditions, in particular about jatropha. This information has encouraged smallholders to adopt crops with great expectations, only to be easily disillusioned when plants need more care, particularly in the initial stages, and provide lower yields. The damage of these initial efforts to create a market waits to be seen.

The manner in which biofuels are being developed is symptomatic of the lack of political influence that poor people have. As was seen in Chapter 3, the rural poor can be completely ignored by the authorities when a large-scale agribusiness moves into an area. The state does little to defend the interests of the poor. However, poor people are not completely powerless to respond to the expansion of biofuel agribusinesses into their communities; for example, Via Campesina,[12] an international movement of poor peasants and smallholders in developing and industrialised countries, vigorously opposes the expansion of biofuels (Borras *et al.*, 2010) while Practical Action,[13] an international NGO embracing technology justice, acts as an intermediary to ensure smallholders get a fair deal (Practical Action Consulting, 2009).

Biofuels, as an energy carrier, are able to address the specific dimensions of poverty: energy and time poverty, which act together to form a vicious circle. Energy poverty is rooted in a limited availability of good quality energy carriers in rural areas. The limited availability is linked to poor infrastructure, attributed to the lack of ability to pay, although it can be argued that it is also due to a lack of political influence by the rural poor.[14] The reasons for the lack of ability to pay are linked to a reliance on traditional energy carriers and time poverty, which is particularly a problem for women. Biofuels as an energy carrier are able to break the vicious circle and bring multiple benefits to households and communities. While electricity dominates the development discourse, and one cannot deny the benefits electricity brings particularly for women in terms of reducing drudgery and improving time management, it is not the best form of energy for many of the end uses that dominate rural households, rural enterprises and communities, requiring either process heat or motive power (see Table 2.1). Biofuels have a distinct advantage over other energy sources: they can provide process heat, motive power and electricity. Biofuel production *in situ* overcomes the distribution problem due to poor roads and difficult terrain faced by petroleum fuels and grid electricity. The literature provides examples of where rural communities are making use of the biofuels that they are producing, in some cases rather than selling to the market. However, there are still barriers to biofuels making a full contribution to reducing rural poverty: some are technological (stoves), some are entrepreneurial (it takes more than a piece of equipment for sustainable income generation) and some are institutional (lack of regulations for microdistilleries in Brazil is considered a barrier to funding for market development (Practical Action Consulting, 2009)).

Another vicious circle which biofuels influences is the link between poverty, health and food security. Poor nutrition can be considered a dimension of poverty which links to ill health. People who are ill are not able to work effectively and sometimes not at all. This means that they may not have sufficient income to buy groceries to supplement the food they are able to grow or harvest from common land. Biofuels can have a positive influence on household food security of the poor by providing income, particularly if women's inclusion is promoted (but at what cost to their own well-being is discussed below in this section). There is evidence to show that food security may also be promoted in biofuel projects by increased output at the household level due to smallholders' access to inputs, such as fertilisers, and improved farming techniques, such as intercropping. However, there is less certainty about the food security of those who rely on common land, when this is designated 'waste or non-productive land' by outsiders and taken over by agribusinesses, resulting in access to ecosystem services being denied, or seasonal wage labour in biofuel plantations, whose wages are low and there is no compensation in the form of garden plots or subsidised groceries.

Biofuels also play a dual role in access to health care. First, by providing the income families can purchase medicines. Second, as energy carriers, they are able to indirectly address core issues of poverty, poor health and illiteracy, by providing process heat and electricity for key community services of health clinics and schools. Whether such services are provided as part of CSR commitment by biofuel developers to communities remains to be seen. Indeed, the literature would indicate that it would be wise not to place too much expectation on CSR delivering such benefits. Primarily CSR was not intended as a development tool to address poverty issues. On the one hand companies, for a variety of reasons, might be motivated to provide infrastructure in areas where they are operating; on the other hand, they may be wary of being open to accusation of assuming the functions of a state without any countervailing powers to moderate any undue influences. Successful development outcomes with CSR seem to be strongly linked to the prevailing local social and political conditions (Newell and Frynas, 2007). CSR is rooted in a business tradition which considers that conflict amounts to differences of opinion, which can be resolved through dialogue. It is, therefore, unlikely that CSR can function effectively where there are significant differences of opinion about models of biofuel production. This can spill over into open conflict and violence, as has been seen in Indonesia and Colombia. CSR has tended to focus on issues related to health and safety rather than human rights (Utting, 2007). Companies with a CSR policy tend to be those that are export oriented and with a high public profile (Blowfield, 2007). This would imply that those programmes oriented towards the domestic market, such as in India, are less likely to provide social services. CSR is considered to be less likely to deliver social benefits in weak states where regulation is not systematically enforced (Dauvergne and Neville, 2010).

If biofuels are to have a considerable impact on rural poverty, there needs to be a significant positive effect on improving women's status and assets. Women form the majority of the rural poor and generally have more limited access and control over assets than men. Nevertheless women are considered to play a key role in household food security and their responsibility for household energy provision. The evidence shows mixed results. As an energy carrier biofuels are able to have a positive effect on household energy providing both process heat and electricity. This provision can do much to reduce poor women's time poverty, improve their health and reduce drudgery, particularly in areas where it is taking several hours for biomass collection. There is also time to be saved in not going to town to buy kerosene and batteries. While there seems to be limited use of biofuels for process heat, it is being used to operate equipment and generate electricity which women are making use of. However, it is quite possible that where biofuel crops are grown on land classified by the state as non-productive, they are actually reducing the availability of biomass, which increases women's time burden, as well as access to other ecosystem services that form important inputs into the household, including food. Women recognise this threat and resist the use of this land. There is no empirical evidence to show whether or not household food security is benefiting from biofuel crop schemes that improve agricultural practices and yields.

Women are participating in biofuel crop production, often more readily than men in the same community. Their incorporation is usually in low income jobs at the bottom end of the value chain, but where their specific lack of resources and skills is recognised and addressed, they are benefiting from incorporation. Women's acceptance of incorporation into the value chain on unfavourable terms may represent their lack of other options compared to men. However, more research is needed to highlight at what cost inclusion is to women's well-being, since the time for participation in BVCs is incorporated into an already long working day requiring management of existing household chores. Another concern is men's reactions to women's access to income, based on evidence from food security programmes that involve women earning income. Men can respond by reducing their input into the household and increase their leisure time. Again we do not know if this is replicated in biofuel production.

While it is difficult to state at the moment the extent to which biofuels are contributing to the reduction of energy poverty and other dimensions of poverty, the potential exists and in some places is being realised. First as a source of income through producing the fuels and second as an energy carrier, which can be used to generate the energy services people require for sustainable livelihoods, in particular improved health and reduced labour and time spent on household activities. In the next chapter we examine how these benefits can be extended and disadvantages reduced.

7   Can biofuels be made pro-poor?

Are biofuel programmes inherently a threat for the rural poor? Does the production of biofuels require specific pro-poor policies to minimise the threats and maximise the opportunities for the poor? Lessons from the early fuel ethanol programme in Brazil would indicate that there is a need for such policies to protect the poor. Sachs has described the early Brazilian fuel ethanol programme as having increased 'concentration of capital, land and power' as well as 'commodifying rural labour' (Sachs, 2007).[1] In other words, it created considerable social inequality: poor working conditions and wage levels for the majority, significant seasonal labour coupled with high levels of internal migration (which in part is urban migration as a result of agricultural mechanisation). How can these conditions be avoided? This chapter suggests answers to this question. The focus is on the institutional aspects of biofuel value chains since these play a key role in ensuring benefits accrue to the poor. To achieve this objective the costs, benefits and interests of the actors in the chain have to coincide. The institutional structure is complex, not only involving different products, different sectors and a range of actors interacting at, and across, different levels.

In Chapter 1, a pro-poor biofuels intervention was defined as one that leads to better outcomes for the poor in terms of improvements in their assets and/ or capabilities. At the very least the rural poor should not move deeper into poverty as a consequence of biofuel production. The poor can be affected, positively or negatively, directly or indirectly, by chain activities even if they do not participate in the chain. How can BVCs be constructed so that the rural poor share in the value added which accrues from supplying an export product? How can pro-poor policies be made to be part of the interests of the powerful? This chapter begins by identifying which production models can be considered pro-poor. It then looks at the role of the governance of BVCs in promoting pro-poor outcomes.

Pro-poor scale of production

The appropriate scale for biofuel production is determined by a number of factors, including the type of feedstock, proximity to markets, the objectives of particular actors in the value chain (for example the state may prioritise rural development through the inclusion of rural people in decentralised energy production systems, whereas a private company may prioritise participation in export markets), the type of biofuel and access to finance. The production chain contains a variety of activities including feedstock production, handling and processing, fuel distribution and marketing which offer opportunities for small-scale producers and can take different forms. Smallholders could produce for a local market and own all parts of the production chain. Alternatively they can grow biofuel crops for sale either on the open market or through contracts to a central processing facility for distribution to more distant (including export) markets.

The processing facility can be owned by the farmers themselves or by private operators (see also next section). In the US, between 1990 and 2004, the volume of production from ethanol processing plants had increased from around 10 million gallons per year to 100 million gallons per year. In 2006, only 26 per cent of new capacity was owned by farmers, which has been linked to the increased size of plants and the difficulties farmers' cooperatives were having in providing the capital to purchase the larger-scale plants and supplying them with sufficient quantities of grain (Kenkel and Holcomb, 2006). This reflects a trend in ownership since the early 2000s when absentee owners began to dominate production again (CFC, 2007).

In Honduras, 80 per cent of palm oil producers have no access to transport and have to rely on middlemen to collect their oilseeds. As a consequence, they have a lower profit margin than growers who have direct links with extractor companies (Fromm, 2007, quoted in CFC, 2007: 57).

The type of crop grown as the biofuels feedstock can favour the inclusion of small-scale growers in BVCs. Many oilseed crops are still harvested by hand which makes the process suitable for pooled smallholder cultivation. Pongamia and jatropha can be grown with intercropping, thus minimising the likelihood of displacing food crops (Raswant *et al.*, 2008). Sweet sorghum is also considered to be appropriate for smallholder cultivation systems and is very similar in properties to grain sorghum grown extensively in sub-Saharan Africa (UN-Energy, 2007). Sweet sorghum's low water demand compared to sugarcane is considered to be pro-poor because poor farmers tend not to be able to afford irrigation (ICRISAT, 2007). Sweet sorghum can be grown in arid areas, such as sub-Saharan Africa. The stalks after sugar extraction can be used as animal fodder, which can be a supplement during the dry season, hence helping to sustain livelihoods.

Biofuel crops can be a new commodity for farmers, who may entail risks and potential gaps in income particularly when waiting for perennial crops to mature, which can take two or three years before there is any harvest. Large-scale agricultural producers have access to insurance and other types of capital to minimise risks. Smallholders tend to be more risk averse so they will need support to shift production systems, particularly when there is a need to wait for a few harvests before the crop yields are sufficient. Project developers do provide support for introducing smallholders to techniques for growing biofuel crops. A unique approach is the establishment by the University of Kasetsart, Thailand, of a jatropha school that trains farmers in oilseed production and processing. The initiative is in collaboration with the Cooperative League of Thailand which provides incentives for its members to grow jatropha as well as guaranteeing prices (Practical Action Consulting, 2009).

Large-scale production systems tend to be favoured for serving export markets. The case for the concentration of production is usually argued on the basis of efficiency related to economies of scale and hence cost (see Table 7.1). When the production chain is vertically integrated the logistics are easier to coordinate as well as the reduced transaction costs associated with negotiations for feedstocks (Kartha *et al.*, 2005). As pointed out in Chapter 3, the logistics of supplying feedstocks by small-scale producers to a central processing plant has proven problematic. Evidence from Indonesia and Honduras, with palm oil production, shows that some small-scale producers struggle to find access to markets (Peskett *et al.*, 2007). In addition, participation in export value chains poses risks for small-scale farmers. There is uncertainty in the sustainability of the market since the demand has been created to meet political objectives linked to energy security and meeting climate change obligations. This means that the market responds not only to the usual market mechanisms but also to political decision making, as can be seen from the European Union's response to negative criticisms of its biofuel policies. There is also the threat from second generation biofuels, which are promoted to avoid the appropriation of land suited to food production.

However, smallholder systems are considered more likely to be pro-poor in their outcomes. To become involved in international value chains, small-scale farmers would benefit from some form of horizontal cooperation (Bolwig *et al.*, 2008). Joining together with other growers brings a number of advantages such as providing a platform for skills and knowledge exchange and learning, a stronger bargaining power than individuals both with other value chain actors and with politicians, as well as being able to provide services to their members, particularly in terms of information, advice and access to credit. From the literature it would appear that farmers' cooperatives are strongly favoured, particularly in Brazil. In the US, farmers' cooperatives have been able to become ethanol refiners. There are examples of where agricultural cooperatives have accumulated assets that have allowed them to start their own research and development directed at the small-scale farmer (ICRISAT, 2007). In these cases, smallholders have been able to move up the value chain and hence gain control of a higher value product. Cooperatives or farmers associations also have advantages for biofuels processors since dealing with a single group reduces transaction costs compared to dealing with a large number of suppliers. On the other hand, there have been some very negative experiences with agricultural cooperatives linked primarily to mismanagement (see for example Hedlund, 1989; Will, 2008) which may make some actors reluctant to adopt this model.

It appears that as biofuel production chains develop, diversification in the numbers of actors involved in the chain begins even when the starting point has been a cooperative (Practical Action Consulting, 2009). Vertical integration is considered in the long term to be inefficient since there is a lack of competition along the chain and the technical efficiency gains which are considered to arise with diversification (Kartha *et al.*, 2005). It is quite possible that the value chain will be dynamic over time, moving between integration and diversification as needs arise and small-farmers' skills and assets accumulate.

The alternative to exporting biofuels is serving local markets (see Chapter 6). This option potentially provides an additional stimulus to rural areas if the biofuel compared to petroleum fuels has a lower price for the consumer and the supply is more reliable than fossil fuels. However, if small-scale farmers supply only local markets they can miss out on benefiting from participating in more lucrative international value chains with the knock-on effect that has for rural development. It may well be in a state's interest to encourage small-holder over large-scale production since such a policy is expected to produce higher returns on public spending owing to greater economic multiplier effects and reduced demand for social welfare due to increased rural incomes (CFC, 2007). If states decide to support decentralised biofuel production to serve local markets, then attention needs to be given to quality control of the fuels to ensure that they meet prescribed standards for use in engines. Failure to do so can not only damage equipment, which in itself is costly for consumers, but it can also damage consumer confidence in the new fuel. Trying to win back trust in a new product can take a long time, as has been found in a number of countries that had introduced biofuels without sufficient attention to the impacts that a change in fuel composition can have on existing equipment (UN-Energy, 2007).

The scale of production may reflect historical processes. In areas where human settlement is well established, then small-scale farming can grow commercial biofuel crops. However, if biofuel production is established in areas where population density is low, it is quite likely that the level of investment required for the infrastructure (such as irrigation or roads) will favour large-scale, vertically integrated production.

Perhaps the scale of production is not so significant for pro-poor outcomes, but rather it is the supporting environment that enables small-scale farmers to participate on favourable terms in both local and international biofuel value chains. The inclusion of the proper support mechanisms can overcome the concerns of the type expressed in Table 7.1 about small-scale production. The supporting environment is discussed below.

Another way for small-scale farmers to benefit from large-scale production schemes is not to grow the feedstock themselves but to rent their land to large-scale producers. There is evidence from Brazil that farmers see this as an attractive alternative to growing feedstock since it represents a more reliable income source (see Chapter 6).

Based on experiences with other agriculture-based value chains it is possible that the most pro-poor gains in terms of increased income are not found on the farm but in the range of ancillary services that develop around

Table 7.1 Perceived advantages and disadvantages of large-scale and small-scale biofuel production systems

Large-scale, extensive production	Small-scale, smallholder production
Generally higher yields and higher efficiency (earnings : cost ratio) due to economies of scale	Generally lower yields and lower efficiency – but note that smallholders can achieve yields and cost savings equal to large-scale
Narrow social benefits	Wide social benefits
Lower and narrower returns on public investment	Higher returns on public investment due to reduced social spending
Lower exposure to risk and greater financial security	Higher exposure to risk and lower financial security
Faster uptake of technology	Slower uptake of technology
Safer private sector investment option	Riskier private sector investment option
Attractive to agribusiness	Attractive to entrepreneurs and small- and medium-scale enterprises
Can supply large-scale biorefineries and commodity export	Can supply local biodiesel and bioethanol refineries as well as larger-scale
More suitable for export-driven policies	More suitable for rural development policies

Source: CFC, 2007: 41

the crop (Will, 2008). As was pointed out in Chapter 2, interventions are needed to ensure that wealthy households, which have the capital to invest in diversification, do not capture all these opportunities.

Governance of the value chain

Small-scale farmers growing biofuel feedstocks are the starting point of the value chain[2] which is a multi-level, multi-actor process leading to the development of a marketable product, in this case bioethanol or biodiesel. The main focus of this book has been on the international biofuel value chain. An idealised example of the value chain is given in Figure 7.1 which shows some of the actors and their linkages. (It is also feasible to construct such a chain for biofuels serving local community markets.) A value chain does not operate in a vacuum. It is embedded in an environment which promotes or hinders the chain's operation and development. This environment has two main components: the macro- and the meso-factors. The macro-factors form an overarching framework and operate at the national and international levels. At the national level there are the formal and informal institutions[3] influencing the operation of the value chain. The formal institutions are the policies and laws in operation, e.g. agricultural support to small-scale farmers and the laws governing land registry. Informal institutions are more difficult to influence, particularly for small-scale producers, such as consumer trends, e.g. the fashion for 'green energy'. There are also a range of institutions that operate at the international level.

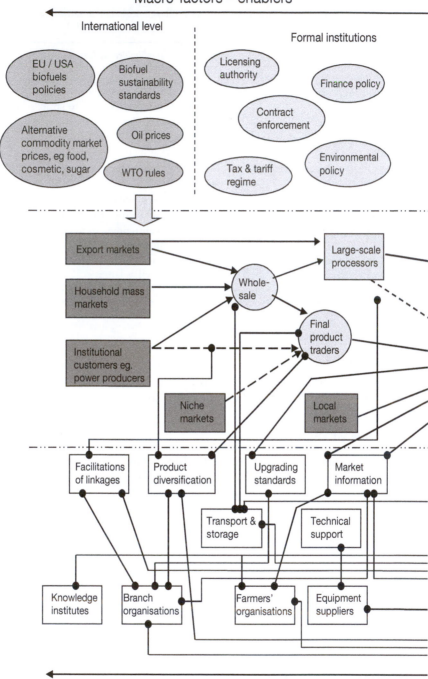

Figure 7.1 Biofuel value chain

Source: author's elaboration

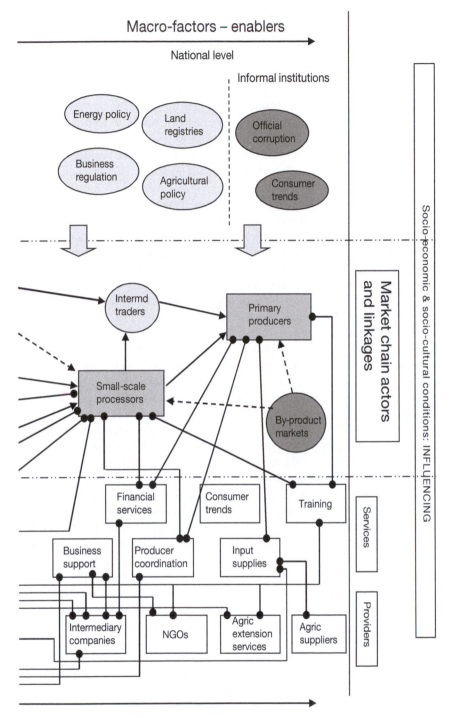

Macro-factors – enablers

National level

Informal institutions

Energy policy

Land registries

Official corruption

Business regulation

Agricultural policy

Consumer trends

Socio-economic & socio-cultural conditions: INFLUENCING

Intermd traders

Primary producers

Market chain actors and linkages

Small-scale processors

By-product markets

Financial services

Consumer trends

Training

Services

Business support

Producer coordination

Input supplies

Intermediary companies

NGOs

Agric extension services

Agric suppliers

Providers

Meso-factors – supporters

The biofuel policies of industrialised countries and the regulations of the WTO are examples of the macro-factors that have significant influence on the biofuel value chain. The macro-level factors can be enablers in the sense that they support small-scale farmers and other rural dwellers to benefit from the biofuel value chain. Indeed, the existing experience with biofuel value chains shows that the regulatory framework plays an important part in the governance (CFC, 2007). However, some of these factors can act as a barrier or even threaten the chain's existence. These factors can also determine the conditions of participation, for example, labour conditions on agri-estates.

The meso-factors support the chain's development. They are the public and private sector organisations that provide the services that help the chain to function efficiently and effectively. They can play an important role in helping small-scale farmers and biofuel micro-producers to participate in the chain, for example, by providing financial and technical information.

The biofuel value chain is new and evolving. It differs from other global value chains in that the primary producers are in one sector (agricultural) but the end of the chain is in another sector (energy). The implications of this structure for small-scale producers have yet to be seen. Experience from agricultural value chains shows that chains are usually constructed in such a manner that they do not favour small-scale farmers in developing countries. There is significant concentration of ownership both in distribution and feedstock trading. Experience with agricultural commodity markets shows that the benefits from export production in the South accrue to actors in the upper part of the chain, while the primary producers receive comparatively little (Dufey, 2007). However, in some cases, biofuel crops (e.g. palm oil and sugar) could also be sold to agricultural chains. Will this give small-scale producers an advantage in negotiations and enable them to command higher prices? Will this cause other actors in the chain to move towards long-term contracts (which may be favourable to small-scale farmers) or to go for vertical integration (which could lock-out small-scale producers) to guarantee supplies? There are signs that even in commodity markets where smallholders predominate, e.g. coffee, large-scale plantations with vertical integration are beginning to emerge (Bolwig *et al.*, 2008). The Cereal Industry Forum considers that the core of governance is likely to shift away from farmers and oil producers and processors towards the downstream end of the chain, that is, the oil and gas chain which is a mature chain with powerful actors (CFC, 2007).

The complexity of the biofuel value chain makes for uncertainty at the government level: where does ministerial responsibility lie? Agriculture? Energy? This uncertainty creates a vacuum which international actors such as agribusinesses, energy companies and investors are able to exploit. These actors have considerably more resources in terms of finance, information, access to legal representation, skills and technology than local actors, particularly small-scale farmers and processors. The political influence on macro-level institutions exercised by the international actors within the value chain is considerable. This can be counteracted by ensuring that there is transparency in transactions through clear statements of organisational responsibility with harmonisation of laws, regulations and directives between different line ministries.

The way in which the multiple actors within biofuel value chains interact will influence whether or not biofuels are pro-poor. However, in global value chains the political interests of powerful actors have to be acknowledged. These actors tend to be located in the upper part of the chain and they can play a key role in controlling the parameters for participation in a value chain by specifying product standards, quantities, timing and price. How this process of control is constructed, that is the governance of the value chain, determines entry to and sets the terms of participation in the value chain. The parameters for participation can be determined by a small number of powerful actors, and when such asymmetrical power relationships exist the more powerful actors can take the larger share of the benefits. Certainly agro–food value chains, which biofuel value chains will probably strongly replicate, are recognised for their highly asymmetrical power relations (Bolwig *et al.*, 2008). So when posing the question, 'How can the biofuel value chain be constructed so that the terms are favourable to small-scale farmers and processors?' we are actually asking, 'How can the interests of the powerful be made to overlap with those of the poor?'

The Lula government in Brazil, aware of the inequalities in power and resources in the biofuel production chain, introduced policies in 2006 that attempt to secure rural livelihoods by using financial instruments which make it attractive to large-scale refineries to source their feedstock from small-scale farmers (Worldwatch Institute, 2007). At the same time the farmers are also supported by agricultural extension services, not only on cash crop production but also on food crops so enhancing food security. Biodiesel produced under this scheme is awarded the 'Social Fuel Seal' and producers are able to use the Seal as a marketing instrument.[4] While there are signs that this programme has brought benefits to small-scale farmers, these benefits appear to have been captured mainly by family farmers in the south and south-east regions, which has been attributed to these farmers having better agronomic conditions, infrastructure and organisation (Bailey, 2008). Emerging criticism from social movements discusses the type of agreements the farmers have entered into with large companies, which effectively keep small-scale farmers at the low value end of the chain with no government support to enable them to own the higher value processing units (Abramovay and Magalhães, 2007). Women in Brazil have also resisted inclusion into biodiesel production. They consider that BVCs changing the scale of production from family farms to large-scale plantations will drive a process of land acquisition, which women will not be able to influence because they do not hold land titles. Therefore, the economic power (in the household sphere) and political power (in the wider society) that they have built will be destroyed (Hospes and Clancy, 2011).

An approach to reduce the risks for the small-scale farmer and processor is the development of long-term contracts with the buyers. These long-term contracts will minimise the risk for the former as well as potentially providing them with improved access to market information, services and inputs (Bolwig *et al.*, 2008). This, of course, supposes that the contractual terms between the small-scale farmer and the biofuel processor are formulated in such a way that neither side is disadvantaged. For example, the sort of situation illustrated by the Coopaf cooperative in Brazil which was contracted to supply 15,000 tonnes of castor beans and only delivered 6,500 tonnes because members sold their beans to middlemen for a more favourable price. The cooperative has learned from the experience and now buys only half of the castor beans grown by members who are then free to sell the other half into alternative markets (Bailey, 2008).

There are models for partnerships between small-scale producers and large-scale enterprises: outgrower schemes, cooperatives, marketing associations, service contracts and joint ventures. Based on the experience with such models from other agricultural commodities, in general it appears that small-holders benefit when there are good contractual arrangements that provide price and purchase security as well as providing support in terms of access to modern technology and research. On the other hand, the private sector seems reluctant to provide farmers with inputs and financial services that are 'attractively priced, timely or reliable' (Dorward *et al.*, 2004: 78) in order to stimulate markets. This is where the enabling environment can to play a role to support small-scale farmers. In particular, the state's role in providing stimulatory policies, at least in the initial stages of biofuels market development, is considered crucial (ICRISAT, 2007). This role is discussed in following section.

There are examples of small-scale producers owning additional parts of the value chain, a process known as vertical integration. For example, in India sugar farmers own shares in sugar mills (ICRISAT, 2007). The government of Mozambique is reported to be supporting a project to develop an ethanol distillery to produce 100,000 litres of ethanol annually based on sorghum grown by 5,000 small-scale farmers (Chege, 2007). Vertical integration can also include taking over the supporting activities such as trading, transport and marketing. Increased vertical integration of the value chain by small-scale producers would potentially increase their bargaining power and influence over value chain governance. Although such a move should not be taken lightly since many of these functions require specialist skills (Will, 2008). This is not to say that small-scale producers should not undertake these activities, only that they will probably require support from actors in the meso- and macro-environments to do so. An alternative way for small-scale producers to have a more equitable share of the value chain benefits is through stronger contracts with downstream actors, which include provisions for support to producers to upgrade their skills so that they increase their efficiency and hence their earnings (Bolwig *et al.*, 2008). The downstream actors need to see such support as creating a win–win situation for example with improved product quality and delivery times.

Another approach for small-scale producers to improve their position in the value chain is through increased collective action as was discussed earlier in the chapter (increased horizontal integration). Collective action can have a number of forms and objectives. Action can focus internally on cooperation between members – for example, an exchange of experiences, which could be very important with a new crop such as jatropha – and providing services, such as bulk purchase of seeds at a favourable price. The action can also be externally focused, for example, in contract negotiation with buyers, and lobbying with actors in the enabling environment for an administrative, legal and infrastructure framework which supports pro-poor development (Will, 2008).

As was pointed out at the beginning of this section, it is not only the actors who participate directly in the value chain but also those who operate in the macro- and meso-environments that can influence the pro-poor nature of the chain. For example, financiers are a group of powerful actors in the supply chain who are concerned about supply risks and return on investments (UN-Energy, 2007). There is little practical experience with commercial production systems for biofuels as well as limited experience in growing a number of the candidate feedstock crops as cash crops, e.g. jatropha. Such uncertainties make financing institutions reluctant to provide support, as do the more general issues related to land tenure as a requirement for loan collateral. Uncertainties surrounding a new crop were certainly found to be the case for a biodiesel project based on palm oil in northern Tanzania when trying to source funds from local banks (Practical Action Consulting, 2009). On the other hand, the international financing agencies are playing an important role in financing investment in biofuels, for example, the Inter-American Development Bank (IDB) is leveraging funds for expanding bioethanol production capacity in Brazil. These agencies are in a strong position to set terms that can benefit small-scale farmers and wage labourers. The IDB has shown a willingness to engage in sustainability issues by developing a Biofuels Sustainability Scorecard based on the sustainability criteria of the Roundtable on Sustainable Biofuels (Amezaga *et al.*, 2010). The Scorecard includes general environmental and social criteria. An alternative source of finance, which has the potential to benefit poor farmers, is the funding from environmental NGOs and ethical banks. Such organisations use sustainability criteria and frequently incorporate social objectives for project assessment. They are also more likely to create access to finance for women, who are often excluded from the formal banking system loans by the requirement of land for collateral.

The current terms of international trade set higher tariffs for agricultural commodities that have been refined into higher value products. For example, the EU charged (as of 2007) 0 per cent and 3.2 per cent respectively on imports of soy bean and crude soy oil for industrial use, while biodiesel imports are charged at 6.5 per cent (CFC, 2007: 51). This type of tariff setting pushes the added value to be gained from processing out of the biofuel feedstock-producing countries into the fuel refining countries, again reinforcing the point that if the processing takes place in the South greater value can accrue to the countries producing the crops. There needs to be clear classification of biofuels as industrial or agricultural goods within the multilateral trading agreement system, otherwise the governance of the value chain is difficult to establish, particularly with any degree of equity (Dufey, 2007).

Biofuels vary in their chemical composition which influences the fuel quality when blended with petroleum fuels, and the environmental impacts when burnt. Therefore, their chemical composition is subject to regulatory standards, for example, the percentage of water permissible in bioethanol to be blended with petrol. High levels of consistency require the use of sophisticated technology, the cost of which may be prohibitive to those without access to financial capital, and the supporting infrastructure has to be in place. Who sets the standards and who verifies the standards has cost implications for producers as well as determining who participates in the value chain and who does not. At the time of writing there is no international regulatory organisation governing biofuels standards, although there are regional initiatives, such as an agreement between the Philippines and Thailand, to set up regional standards for bioethanol (UN-Energy, 2007). Therefore, exporters to multiple markets may face having to meet a range of different standards and the associated costs of compliance. If the cost of compliance standards becomes prohibitive, it could shift the balance of processing feedstocks from the South to the North and hence the loss of the value added component of the supply chain. Indeed, the technical specifications for biofuels can become a significant barrier to participation in international BVCs by small-scale producers.

There has been considerable discussion about mechanisms for increased democratisation of the value chain to counteract asymmetries of power in order to bring greater pro-poor benefits. How actually to do this is a challenge given that global value chains operate on market principles and it is unlikely that all members of a chain, or indeed their representatives, would sit around a table together as a matter of course. A possible starting point for influencing the governance of biofuel value chains is in the planning processes for biofuel projects which could incorporate a greater participation by all actors. A neutral forum that brings actors together can create transparency and avoid the types of misinterpretation described in Chapter 3. Such a forum opens up the potential for the rural poor to participate. Keeping in mind that the rural poor are not a homogeneous group care needs to be taken that the forum is representative of all stakeholder groups. NGOs and CBOs can play an important role here in creating a level playing field and ensuring that there is a fair deal for the rural poor. An evaluation of 15 small-scale bioenergy projects[5] spread across the South clearly showed that when there was a leading role for NGOs and CBOs in a bioenergy project there was considerable attention on equity[6] (Practical Action Consulting, 2009).

The establishment of a global regulatory body (or bodies) for biofuel standards would oversee fuel quality and sustainability but such organisations would need to guard against imposing conditions that are so costly to comply with that they undermine the efforts by the rural poor to participate in global value chains. In terms of fuel quality, there are concerns that the cost of meeting these standards will reduce small-scale farmers to growing only the feedstock and they will lose the possibility to participate in the value added component of crop processing. The costs of compliance to small-holders have been estimated at about 20 per cent of production costs (Cramer Commission, 2007). Group certification might be a solution to keeping costs down for small-scale producers.

Ensuring compliance with biofuel sustainability standards can be scientifically more challenging than for other commodities, such as coffee and timber, because the product is not sold as a distinct entity that can be easily labelled (Verdonk *et al.*, 2007). Indeed, the challenge seems to have been beyond the resources of the multinational Unilever, which admitted in 2008 that it was not able to trace the origin of palm oil supplied by firms operating in Indonesia (Greenpeace, 2008). However, sustainability is generally considered to be multidimensional and hence will address more than the social dimension. In terms of environmental sustainability, small-scale farmers are not always 'green' and actors in the downstream components of the value chain may be more responsive to consumer demands for environmentally friendly production practices (Reardon and Barrett, 2000). However, there are doubts about whether or not consumers will be sufficiently enthusiastic for 'green biofuels' to drive the process, as they are for sustainable food and timber products (Verdonk *et al.*, 2007). On the other hand, there are a number of issues with strong consumer interest that form a nexus around biofuels: climate change, genetically modified organisms, dependence on oil imports and impacts of development paths. How these issues will be optimised by consumers and translated into their behaviour at the petrol pump remains to be seen but their behaviour as consumers can certainly influence the economics of biofuels and apply pressure for more pro-poor biofuels.

There are some actors, particularly NGOs, who see the certification schemes that exist for a number of commodities (for example, timber and the Forestry Stewardship Scheme)[7] as a template for getting biofuel value chain actors 'around the table' (UNCTAD, 2008). However, these schemes are about addressing a broad range of sustainability concerns rather than specifically ensuring a fairer distribution of financial benefits to small-scale producers. Therefore, these schemes would be seen as addressing more than the economic dimension of poverty, for example, ensuring respect for legal and customary rights of indigenous people.

A number of sustainability initiatives with respect to biofuels exist. These initiatives can be put into two general categories: multi-stakeholder initiatives, such as the roundtables on sustainable palm oil and soy; and government-led schemes such as the EU's proposed biofuels sustainability criteria. The former combines a roundtable deliberation process with the development of a set of voluntary sustainability criteria as well as an internal governance system (Cotula *et al.*, 2008). The schemes are organised by a variety of actors: importing country governments (e.g. UK and the Netherlands), NGOs (e.g. WWF), international agencies (e.g. FAO) and multi-actor platforms (e.g. Roundtable on Sustainable Biofuels). Yet a proliferation of schemes creates uncertainty, which could deter small-scale producers from participation particularly since the cost of verifying compliance with the criteria has to be borne by the producer. It is claimed that in 2008 the costs of developing, documenting and certifying Environmental Management Systems to comply with RSPO (Roundtable on Sustainable Palm Oil) accreditation was 'in the range of five to ten per cent of the company's crude palm oil production cost' which was discouraging small- and medium-sized companies from seeking certification (*New Straits Times* 2008,[8] quoted in McCarthy and Zen, 2010). An additional deterrent is the large markets, such as Indonesia, India, and China, where environmentally aware consumerism is nascent and vigorous regulators are scarce.

The state-led initiatives serve as a policy tool for discriminating against sustainable and non-sustainable production systems. Whether these schemes could be used as criteria for import restriction, e.g. into the EU, to support pro-poor policies is questionable since such an action might be ruled illegal under the WTO laws of restricting trade (Keam and McCormick, 2008). The EU decided to omit social criteria from its standards to prevent any possible conflict with the WTO rules (UNCTAD, 2008).

Interestingly these initiatives have been welcomed by some downstream actors as a means of improving the image of biofuels (Colchester *et al.*, 2006). Companies from both downstream potentially competing chains (agriculture and oil) are members of biofuel roundtables, for example, Cargill, one of the world's largest traders of grains, and the oil giant, Shell. Shell has appointed a biofuels compliance officer, with a support team, to ensure that Shell sources only biofuels from sustainable sources, which includes clauses in its subcontracts that are intended to ensure that there are no human rights violations (including child/forced labour) in the production of the biofuels (UNCTAD, 2008). It remains to be seen how effective the powerful actors in the value chain are at influencing the social and environmental aspects compared to the more technical aspects of compliance with chemical standards.

Certification schemes have been criticised for the lack of clarity in definition of criteria, for example, 'workers should not be unnecessarily exposed to hazardous substances' and 'there should be equitable landownership' (Lewandowski and Faaij, 2006). The Roundtable on Sustainable Palm Oil has a certification process that covers land rights, labour laws and biodiversity protection (RSPO, 2006). However, many of these criteria are simply a reiteration of existing laws on employment regulations and protection of listed species, and so a producer would be expected to comply with them in the course of normal operating practices. Some of the criteria are quite complex. For example, the Benchmark of the Better Sugarcane Initiative Principles and Criteria of November 2009 has 5 principles, 20 criteria and 46 indicators covering requirements to obey the law, respect human rights and labour standards and manage biodiversity and ecosystem services (Lovett *et al.*, 2011). While the principles that underlie the criteria are well-meaning, vagueness in formulation of criteria results in indicators that are problematic to quantify, which can make it difficult to hold powerful actors to account – the very thing the schemes are intended to do. A review of certification schemes in 2006 found that there were no schemes using indicators for food and energy security, local benefits of biomass trade and abatement of poverty (Lewandowski and Faaij, 2006).

At the moment the schemes are mainly Northern driven which raises not only issues of equity of participation but also the sensitivity to local environmental and social conditions (CFC, 2007; Verdonk *et al.*, 2007). The early signs are not positive; for example, the Roundtable on Sustainable Biofuels has been criticised for drawing up standards without consulting NGOs speaking on behalf of smallholders from the South[9] (Ernsting, 2007). The schemes have also been criticised for their 'one-size-fits-all' approach and failing to tailor solutions to local conditions. 'Some schemes may favour particular process and production technologies that may be unavailable, unsuitable or prohibitively expensive for trading partners' (UNCTAD, 2008: 29). Where participation of small-scale producers in meetings to develop sustainability criteria is promoted, there is a need to guard against tokenism and giving legitimacy to decisions that rural people do not support. Multiple–actor roundtable meetings are not necessarily discussions among equals and decisions can reflect existing vested power relations.

At the time of writing, Brazil is the only country from the South to have started to develop its own national certification scheme for biofuels, which aims to make biofuel production conform to environmental, social and labour standards. The scheme will include compliance with environmental and labour laws, adequate work conditions and the socio-economic development of the areas surrounding the production fields (UNCTAD, 2008).

The policy context

Agricultural economists have long considered agriculture as an engine of growth in the early stages of development. Growth in the rural economy, based on agricultural growth, is considered to have a stronger effect on poverty reduction than growth in other sectors including manufacturing (Datt and Ravallion, 1998, quoted in FAO, 2008a: 80). Biofuels are, therefore, considered as providing the opportunity to stimulate the rural economy, to generate growth and reduce rural poverty. However, if poverty reduction is to be achieved through increases in direct household income, then attention needs to be paid to land distribution particularly for women, because equitable land distribution is considered to give a more equitable distribution of the benefits accruing from agricultural growth (Lopez, 2007, quoted in FAO 2008a: 80).

The pro-poor nature of biofuels programmes can be considered at its best when the rural poor are direct beneficiaries and at a minimum when the poor are not moved deeper into poverty as a consequence of biofuel production. In the South the state is a key actor for ensuring that biofuels are pro-poor.

The state creates the policy context which shapes the form of any biofuels programme. The policy instruments the state uses play a significant role in determining which actors participating in the biofuel value chain will benefit from any support provided. The state can make interventions directly to support specific sectors linked to biofuels, such as agriculture, or indirectly, for example, by generally stimulating foreign direct investment.

The biofuel value chain can be considered to have three main components: production, processing and consumption. Where the interventions in the chain are made is also crucial in ensuring pro-poor outcomes.

Low income farmers may be expected to directly participate in the agricultural production stage of the value chain. The state may provide support to agriculture, such as subsidies on inputs and provide appropriate advisory services and credit mechanisms. The point along the value chain to target subsidies has been the subject of much discussion. In the US, support to growers rather than blenders is considered the most appropriate instrument to help rural areas as the former are more likely to be rural based. Studies in Minnesota and Iowa indicate that locally owned bioethanol plants return US$0.75 for every US$1 in local activity, compared to US$0.25 for absentee owned plants (CFC, 2007). Using the volume of production of individual producers as a limiting factor for receiving financial support is also seen as a way of targeting support to small-scale farmers.

As was pointed out in Chapter 3, if biofuels are to have pro-poor outcomes it is not only the instrumental factors that need to be addressed but also the political issues, such as land distribution and tenure. The state can also be pro-poor by selective support for the types of biofuel crops grown. For example, in Brazil, soybeans are predominantly grown by agribusinesses rather than family farmers, so a pro-poor policy choice would be either to not include soybeans for support, in this case under the PNPB, or to devise a formula only allowing farmers to qualify for a quantity of beans commensurate with the output from a small farm.

An indication of whether or not the rural poor are benefiting from targeted interventions designed to stimulate the rural economy can be gained from an analysis of Poverty Reduction Strategies (PRSs) which have had a significant influence on the development of pro-poor policies in the South. An analysis of PRSs has shown that the policies and interventions tend to have a bias towards spending on health and education (strengthened by a linkage with the Millennium Development Goals) and neglecting the rural productive sectors (Cromwell *et al.*, 2005). In part this can be attributed to the lack of consensus on whether agriculture can be the engine of pro-poor growth and the role of small-scale farmers in economic growth (Ashley and Maxwell, 2001). At the policy level, there appears to be only loose linkages between economic growth in the agricultural sector and poverty reduction. As a consequence many policy makers do not consider investment in agriculture to be the most appropriate mechanism for poverty reduction (Dorward *et al.*, 2004).

At the same time energy policy is also not taking a pro-poor approach. Again taking PRS papers as an indicator of pro-poor goals, the papers do contain references to the significance of energy in national economic development although elaboration of energy poverty linkages is not widespread (UNDP, 2006), despite the energy poverty linkages being recognised for some time (see for example Barnett, 2000; Reddy, 2000) as well as the gender dimension (see for example Clancy et al., 2003; Ramani and Heijndermans, 2003). An analysis of the 54 PRS papers available in mid-2005 found only one with a distinct reference to energy for poverty reduction: Mali. Interestingly the reference is the biofuels project described in Chapter 6, which has been designed to reduce women's time and drudgery burden linked to increasing the profitability of agricultural production (Burn and Coche, 2001). When it comes to making pro-poor investment choices in the productive sector, there is a lack of understanding about how the poor make a living which leads to inappropriate policy decisions (CPRC, 2008).

It was indicated above that the role of the state in creating pro-poor markets is pivotal. However, the exact nature of that role is contentious since there are many examples of failures of state interventions in providing services to small-scale farmers, and there are fears that subsidies can become unsustainable fiscal burdens. However, there are examples of states taking a pro-active stance to ensure small-scale farmer participation in growing biofuel crops. For example, to ensure the participation of small-scale farmers in biodiesel production, the State of Chhattisgarh, in India, has distributed 380 million jatropha seedlings free of charge as well as oil presses to *panchayats*[10] and a guarantee to buy seeds (Fairless, 2007). Another example of state intervention is the Social Fuel Seal programme in Brazil which encourages biodiesel producers, by the provision of tax exemptions, to purchase feedstocks from family farms in poorer regions. By the end of 2007, it was reported that approximately 400,000 smallholders were participating in the scheme (FAO, 2008a: 83).

As was pointed out in Chapter 3, there is considerable value added in the processing stage and so creating opportunities for small-scale farmers to participate could be supported by policies, for example, through sponsoring innovative approaches to rural finance (FAO, 2008a) that are amenable to small-scale farmers. Existing capital markets are considered unfavourable to small-scale investments (Amigun et al., 2008). Here the state can play a role in providing the right type of information that can remove the barrier to finance, while third parties, such as NGOs, can play a facilitation role.

Biofuels are a new product that requires the creation of distribution channels, not only involving good infrastructure, such as roads and communications, but also an integrated institutional structure. In Ethiopia, the Action Plan to Blend Ethanol with Gasoline gave a target mixture of 5 per cent ethanol/75 per cent gasoline by 2007. As of 2008, no bioethanol fuel blends were on sale. This has been attributed to a failure by the government to ensure that all elements of infrastructure for storage, blending, transport and distribution were in place (Mulat, 2008). How standards, blending ratios and targets are arrived at can be crucial to their success. In the biofuel programme in Sweden, rather than imposing blending targets and distribution mechanisms, standards were negotiated by a public foundation with oil companies, car manufacturers and other stakeholders. The process was considered successful in ensuring that social goals were paramount to profit goals (Kartha *et al.*, 2005). A model for overcoming the type of impasse in Ethiopia is to build the supply chain step-wise by starting at the end with the consumer and working back along the chain finishing with the grower. This approach was used in Sweden where imported ethanol was used to fill the gap until there was sufficient local production (Kartha *et al.*, 2005). There are doubts about whether or not some countries have the capacity to deal with the complexities of biofuel projects. In Tanzania, one company found it took a year to register the land for their biodiesel project. There was only one government official authorised to carry out the registration and their services were in considerable demand (Practical Action Consulting, 2009). The capacity of state institutions to help track land deals has been questioned and official statistics are often out of date (Vermeulen and Cotula, 2010)

The role of state institutions at the local level in the biofuel value chain tends to be overlooked. In Mali, the local administrations have set up village producers committees which serve as the distribution node for jatropha seeds (Practical Action Consulting, 2009). Depending on the context, because their legal powers and responsibilities will vary from country to country, local administrations may play a role in land allocation for biofuel project development or this can be the sole responsibility of the central state. Whether they have the capacity to ensure that villagers within their jurisdiction are not disadvantaged by powerful political and economic forces remains to be seen. Local government officials are better placed than central ministries to assess if land is 'waste'.

In the consumption stage of the value chain, setting targets for the blending of biofuels with petroleum fuels is considered a key driver of biofuels programmes (FAO, 2008a). However, the way in which targets are set by the state can be problematic for small-scale farmers who are not always able to respond as quickly as large-scale producers in terms of investment decisions and organising the necessary inputs such as finance. The state can adjust policies quickly and create uncertainties. In 2007 in Brazil, for example, the government announced a change in the target date from 2013 to 2010 for a blended fuel of 5 per cent biodiesel and 95 per cent fossil diesel (Bailey, 2008).

Biofuels are considered not to be economically competitive with fossil fuels without subsidies, even when the crude oil price was at the high of more than US$100 per barrel experienced in the 1980s.[11] Therefore, states face difficult investment decisions, in particular where in the value chain to provide financial support. In terms of benefiting the poor, research shows that investment, particularly for roads and other infrastructure in regions where the poor live, is considered to have a high poverty reduction impact (Evans, 2007, quoted in CPRC, 2008: 28). Local agricultural research and extension services are also considered important factors in supporting poor farmers. Therefore, the establishment of biofuel production in regions where the poor live and investing in infrastructure in those regions could potentially bring multiple benefits to the poor, including those not directly involved in the BVC. An alternative to subsidies is to send clear signals to the market, such as mandatory blending requirements, which creates investor confidence (Kartha *et al.*, 2005).

There have been doubts raised as to whether or not an enabling environment (e.g. infrastructure, extension services and credit market) to support and stimulate investment (particularly foreign investment) in biofuel production exists in many developing countries (Kojima and Johnson, 2005). Creating the enabling environment for investment is a critical role for the state, including a 'clear, stable and transparent legal and fiscal framework' (CFC, 2007: 49). Policies which promote biofuels specifically or as part of a renewable energy portfolio have been considered to provide a major boost in the North, for example, in the US and the EU (Amigun *et al.*, 2008). However, such policies are lacking in many countries in the South. The most notable exception is Brazil, where the state in the 1970s took the strategic decision to develop a bioethanol programme partly in response to the oil price rises and the impact that these had on the import bill.

On the other hand, there are questions raised about the capacity of state institutions to deal with the levels of foreign direct investment that can be involved in large-scale biofuels development. For example, between 2007 and 2009, over US$1 billion was invested in five African states (Dauvergne and Neville, 2010). There needs to be an established autonomous state bureaucratic structure that is capable of overseeing these levels of investment, with the appropriate legal and fiscal frameworks to regulate and maintain control of local production to ensure that benefits accrue to the local economy. The concern, based on experiences in the agrifood business, is that where these structures are weak, transnational corporations will be able to capture the benefits of biofuels at the expense of local communities and the national economy (Dauvergne and Neville, 2010).

So it can be concluded that the policy environment as presently constructed would promote biofuels as part of a pro-growth strategy rather than a specifically pro-poor strategy. The poor, it is assumed in a pro-growth strategy, will benefit through a trickle-down effect (CPRC, 2008). There is a lack of policy coherence at the national level, which is made more complex because the biofuel value chain cuts across sectors and biofuels programmes are being formulated to meet multiple goals, again arising from different sectors, such as fuel security and poverty reduction. For states in the South with policy aims of serving export markets, there is a considerable uncertainty surrounding biofuels over which they have very little control making investment decisions risky. There can also be weak enforcement of policy, including protecting biodiversity and ecosystems, which sends mixed signals to actors in the value chain.

Biofuels are only one possible option for transport fuels. While it is true that they are an attractive solution for petroleum products substitution since they can use the same delivery infrastructure and the existing vehicle fleet, they are not the only option. Decisions taken in the industrialised economies, and the large emerging markets of China, India and Brazil, can provide an impetus to biofuel production or kill it by their policy decision about the choice of transport fuel. Similarly, use of second or third generation biofuel technologies can result in industrialised countries being able to meet most of their liquid fuel needs and render Southern exports surplus to requirements. This is an argument for biofuels being promoted to serve both local and export markets rather than focusing entirely on the latter.

Ensuring the rural poor a fair deal

The underlying premise in this book is that the rural poor, whether they participate in BVCs or not, should at the very least not be moved further into poverty as a consequence of biofuel production in their locale. How can we ensure a fair deal for the rural poor? Rural people are not a homogeneous group (see Chapter 2). People have multiple identities differentiated along lines of status, wealth, gender, age, ethnicity, clan and occupation. Such a differentiation leads to the emergence of groupings with divergent views on biofuels which will be localised. Some groups may oppose the development of biofuels in their communities while others (often local elites) may welcome the opportunity and seek to exploit their position. For example, customary chiefs can enter into deals with biofuel developers to the detriment of other local groups, as happened in West Kalimantan, Indonesia (Cotula *et al.*, 2008).

Farmers can vary in their views about biofuels. The International Federation of Agricultural Producers (IFAP) considers that:

> Bioenergy represents a good opportunity to boost rural economies and reduce poverty, provided this production complies with sustainability criteria. Sustainable biofuel production by family farmers is not a threat to food production. It is an opportunity to achieve profitability and to revive rural communities.

While Via Campesina considers that:

> the social and ecological impacts of agrofuel development will be devastating … They drive family farmers, men and women, off their land. While TNCs [transnational corporations] and investment funds increase their profits, a large part of the world population does not have enough money to buy food.
> (Borras *et al.*, 2010)

The IFAP is an organisation composed of commercially oriented small-, medium-sized and rich farmers, whereas Via Campesina is an international movement of poor peasants and small-scale farmers in the developing and industrialised worlds.

At the local level, there needs to be the political will to ensure that the rural poor do not become poorer from biofuel programmes. Reducing the vulnerability of the poor by addressing issues such as lack of land title and ensuring a respect for customary rights with regard to common land could help them to resist local elites and powerful external commercial forces as well as enabling better land management, so protecting natural resources and increasing productivity (Prowse and Chihowu, 2007). Indeed, who is able to own land may become a determining factor in who can participate in global value chains. Reardon and Barrett (2000) report that changes in Peru in the 1990s enabled agro-industrial firms to own land, which has induced vertical integration by firms and land acquisition by processors to the detriment of local farmers.

Schemes based on small-scale producers to supply a large-scale production plant could be a strategy to ensure farm holdings are not lost. Institutional arrangements should ensure that smallholders are not disadvantaged both at the global level through the international value chain, and at the local level in terms of their contracts with local biofuel refiners.

Giving rural people a 'fair deal' from biofuels is not only about the terms of inclusion in BVCs but also about the nature of the land allocated to outsiders and how this is used. The evidence in Chapter 4 shows how land-use changes threaten ecosystem services, which form an essential contribution to household daily survival and form the basis of livelihoods as well as acting as insurance in times of hardship. Although threats to ecosystem services from LUC changes to forested and crop land are more tangible it is the use of land classified by outsiders as 'waste', 'under-utilised' or 'degraded' where the services are less obvious, and poses a significant threat to the rural poor.

One of the solutions to ensuring that the poor are not disadvantaged by biofuel programmes is rather glaringly obvious: enforcement of laws and regulations that protect the interests of the poor. However, often the political will is lacking to protect the poor and marginalised. For example, the displacement of poor people by biofuels programmes can take place when customary practice is ignored by formal legal systems. However, this is not a problem exclusive to biofuels. On the positive side, there are examples of government initiatives to ensure that landless people can participate in biofuel crop production. The state government of Andhra Pradesh in India has helped landless people set up self-help groups and then granted them usufruct rights to harvest trees planted on degraded common land without transferring land rights to the groups (ICRISAT, 2007).

Paying compensation for loss of access to land and other resources is also an option which does not necessarily have to be by direct cash payments since cash transactions in weak financial markets, such as a lack of banking services, can be problematic. Instead alternative forms of compensation have been promoted, for example, investment in infrastructure such as irrigation that could benefit farmers (Vermeulen and Cotula, 2010). Such approaches may be considered more transparent, because it is visible where the benefits accrue, whereas financial compensation to 'communities' may be less so.

For the landless or those dispossessed of land, biofuels bring hope of employment. Indeed, biofuels are promoted for the stimulus to the rural economy including creating jobs for the landless (Chapter 1). As shown in Chapter 3, the numbers of jobs created and the skills level have been disappointing. An analysis of 150 Ethiopian land deals related to biofuels found that 130 offered fewer than 50 full-time equivalent jobs mainly requiring low level of skills (Vermeulen and Cotula, 2010). Some estimates give the number of jobs per 100ha generated by plantations as one-tenth the number of jobs generated by family farming (Borras *et al.*, 2010).

The evidence suggests that those who are employed as plantation workers are the group whose situation does not improve or even deteriorates as a consequence of participation in BVCs. Once established, large plantation production systems do not have significant impacts on rural unemployment. Jobs that exist are generally low skilled and seasonal. There is a push to mechanisation, often under the guise of a response to international condemnation of conditions endured by manual workers both in terms of employment (hours, health and safety, methods of payment) and in terms of living conditions. The use of gang-masters and subcontractors for supplying plantation workers, which continues to be problematic in Latin America, is closely linked to the use of forced and bonded labour. Such workers often struggle to oppose their working conditions because they are denied their right to organise and bargain collectively.[12] However, this exploitation of agricultural labour is nothing inherent in biofuels but in the system of agricultural production and asymmetries of power.

The freedom of association and collective bargaining are often included in corporate social responsibility (CSR) standards. Disadvantages of these standards are that they rely on self-regulation, the capacity for monitoring is often lacking, particularly in the South (Blowfield, 2007) and companies learn how to get round standards (Prieto-Carrón *et al.*, 2006). In Chapter 6, it was pointed out that relying on the ethical behaviour of companies, such as through CSR initiatives, may be limited to companies with an international reputation to protect. Ethical approaches seem to have escaped a number of companies in Latin America operating in domestic markets where the exploitative practices of labour continue. However, legal uncertainty remains about whether or not WTO and GATT rules allow import regulations that would discriminate between producers and their compliance with certain labour standards. This is one of the reasons that the EU has not included social criteria in its sustainability standards for biofuels (UNCTAD, 2008).

Rural women are another group who have been identified as threatened by large-scale biofuel production (Rossi and Lambrou, 2008). In Chapter 3 it was shown that the threats to women's livelihoods from biofuels are no different to threats from any other cash crop. Enabling poor women to benefit from biofuel programmes requires institutional change. Women's ownership and control over land and other forms of property need to be enshrined in law and these rights need to be enforced. While it is well known that ownership of assets opens access to credit, it is probably less well recognised that women's ownership of land, 'even a small piece'[13] (Kelkar and Nathan, 2005: 26) can have a positive effect on women's bargaining power in the labour market, which is not observed when the land is owned by men. In other words, when women have no alternative (their labour is not needed for harvesting crops on their household's land) they can be reduced to working for lower wage rates than men. However, ownership of land alters the bargaining power for women and they can demand higher wage rates because they have an alternative. If, through higher wage rates, the opportunity cost of women's time increases then the adoption of labour-saving equipment is likely to be promoted. Therefore, the workload of women does not necessarily show an accumulated increase through participation in labour markets if the time taken for their household tasks is reduced by the labour-saving equipment. It is also possible that women's time can be freed to participate in labour markets if within the household there is a transformation in the gender division of labour, with men being prepared to do some household tasks, an effect that has been observed after rural electrification allowing the use of electric equipment (Clancy *et al.*, 2006).

As stated above, women are not passive actors in biofuels development but will resist inclusion in BVCs if they consider the terms detrimental to their existing status, or they will opt for inclusion if they consider this opens up new opportunities not already colonised as a 'man's domain'. There are signs that where gender equity is taken into account during project design, women can benefit from biofuel projects. India has a number of positive examples where women's self-help groups have clearly defined roles both in the governance of the value chain and the production of biofuels (Narayanaswamy *et al.*, 2009; Practical Action Consulting, 2009).

Food security

The issue of food security is an aspect of the biofuels debate that has aroused international attention and has considerably polarised attitudes towards biofuels. As outlined in Chapter 5, there has been much well-meaning but ill-informed comment on the effects of biofuels on food security of the poor. However, while we consider it essential to monitor LUCs and their potential impact on food production that might occur if biofuel crops are under consideration, we would suggest that too much focus on growing biofuels as the identified problem detracts from the underlying causes of food insecurity and hunger. Food security has been a concern long before the emergence of biofuels and a complete ban on growing biofuels will not solve the problems of hunger. Indeed, as history shows,[14] a food surplus will also not end hunger. As pointed out earlier, large quantities of food go to waste due to the lack of infrastructure to process, store and distribute food. Certainly there can be increases in output particularly in Africa where investment in agriculture remains seriously deficient. Investment in postharvest processing, storage and transport infrastructure is needed to increase food availability, *irrespective of what happens with biofuels.* Energy is an important input for increasing food availability and locally produced biofuels can make a contribution as an energy source, in other words as part of the solution of food insecurity (Chapter 6).

Biofuels are not the only energy-related threat to food security: climate change is considered a major threat to crop production. Although it is still difficult to make precise predictions about the impact on crop yields of the changing rainfall patterns (levels and frequency) and higher temperatures, indications of yield declines for the most important crops are apparent (Nelson *et al.*, 2009). Not only will this affect food prices, which will be detrimental for both the urban and rural poor, smallholder income is also expected to decline due to reduced levels of output. The most negative impacts are likely to be felt in areas where food production is already deficient. It has been estimated that agricultural production may fall by 30 to 40 per cent in India and 20 per cent or more in Africa and Latin America by 2080. Sub-Saharan Africa is predicted to face a decline in food production of more than 50 per cent (Cline, 2004, cited in Memedovic and Shepherd, 2009: 32).

There is also the trend of rich, food deficit countries gaining access to land, most notably but not exclusively in Africa, to produce for their domestic market rather than the 'host' country's population. There are LUC effects, similar to those described in Chapters 4 and 5 for biofuels, when large-scale food crop plantations begin to compete with local smallholders for resources: most notably water. Where water is in short supply, conflicts related to access may arise and can impact on levels of output particularly for smallholders whose rights are not respected. On the other hand, it is suggested that inward investment can improve output of all producers through access to capital, technology, know-how, markets and improved infrastructure (World Bank, 2007). Land deals are not always available in public records, therefore figures available for the quantities of land involved are probably an underestimate; but for Africa the range is from 0.6 per cent in Mali to 2.3 per cent in Madagascar (Cotula and Vermeulen, 2009: 1241). However, it is probably not the scale of land area involved but the quality of land and its existing uses that are potentially problematic.

However, this is not to advocate complacency about biofuels and food security. The attention to the issue of food security which biofuels brings to international debates might provide the pressure on states to take action. In order to safeguard food security, in the context of large-scale agribusiness biofuels (and other cash crops), project developers need to take a conscious decision that this will be an important objective in their projects. It can be backed up by state regulation and also by industry standards, although, as pointed out in the previous section, companies can be lax in implementing ethical standards. There are positive examples of biofuel project development where food security is an integral part of the project design. For example, in Tanzania, a company which buys palm oil fruits from outgrowers only purchases fruits above a certain size and the remainder are left for serving the local cooking fuel markets (Practical Action Consulting, 2009). As a note of caution, it appears from the literature that the examples of such conscious design tend to have significant NGO or CBO involvement. It remains to be seen if large agribusinesses adopt such schemes if left to their own devices.

Biofuels are a threat to food security through the LUCs when 'waste land' is used by outsiders for biofuel crops. Land that is not currently producing commercial crops is a source of food for the rural poor, particularly the landless, and for grazing small animals. Such ecosystem services can also be a source of income for the rural poor enabling them to earn income to purchase food. It is these aspects of food security that are less well understood and receive less attention.

Concluding remarks

Biofuels are currently being promoted as a solution for fuel security, high oil prices, climate change, rural development and securing new markets for farmers. However, this is a technological fix for what are complex social, economic, political and ecological problems. The apparent contradictions have been summed up by Achim Steiner, UN Under-Secretary General and Executive Director, UNEP (United Nations Environment Programme), in his introduction to the 2008 report of the German Advisory Council on Global Change. 'Biofuels have been represented by some as a silver bullet to the climate change threat, and by others as a fatal mistake set to destroy forests and increase hunger; they are neither' (German Advisory Council on Global Change, 2008).

There is limited empirical evidence to make a definite statement about the effects of biofuels on rural people because biofuels are currently not as widely produced as commercial products. Therefore, there is a tendency to generalise about the impacts of biofuels in the literature, particularly in the non–academic literature, based on the experiences with bioethanol from sugarcane in Brazil in the 1970s and 1980s. Such generalisation is inherently dangerous, and the analysis tends to neglect the context. The northeast of Brazil is a region with a long history of conflict between the poor and the large landowners. The bioethanol programme of the 1970s and 1980s was the backdrop to political power struggles taking place during a period when Brazil was governed by a junta with strong links to ruling elites in northeast Brazil (M. Lehtonen, personal communication). These power struggles continue today. However, it would be incorrect to attach the social, sometimes violent, transformations that took place in Brazil as something inherent in biofuels. That is not to say that such power struggles will not take place within biofuel programmes, as can be seen in Colombia and Indonesia, but it needs to be recognised that these struggles are a reflection of much broader processes within a specific context, which require a deeper analysis (Reardon and Barrett, 2000).

The type of simplistic criticism of biofuels that seems to dominate much of the policy discussion (such as labelling them 'a crime against humanity') is not entirely beneficial to the rural poor in the South, in fact it runs the risk of disadvantaging them. Biofuels do offer the opportunity for the rural South to participate in new markets. There is a difference between biofuels that allow for product diversification by actors already participating in commodity markets (for example, the sugarcane producers of the Caribbean and tobacco growers of Malawi) and those that require the establishment of new biofuel plantations. In the former, there is a need to ensure that any existing inequalities are not perpetuated, for example, that there is a more equitable distribution of benefits and an improvement in working conditions. When establishing new biofuel programmes, the speed at which the establishment of such schemes may be particularly damaging to the interests of the poor. Again the damage takes different forms and hence requires different solutions, for example, customary land rights may be ignored, production chains are established that do not create opportunities for small-scale farmers to participate and feedstock crops are chosen that have consequences in terms negative to environmental impacts and the numbers and types of job opportunities created for the rural poor.

Demonising a particular crop, most notably jatropha, also misses the point. What matters for pro-poor outcomes is which crop, grown under what agronomic conditions, under which forms of ownership and labour regimes and the construction and governance of the value chain. There is nothing inherent in a chain based on foreign capital invested in large-scale plantations using monocultures as the only possible way to produce biofuels on a large-scale. Such a choice is an economic decision as part of a strategy used by groups to pursue a particular agenda supported by the policy framework (in other words a political decision) (Hospes and Clancy, 2011). What is a concern is that these policy frameworks in the South are being influenced by Northern companies to promote one type of chain to the exclusion of others (Borras et al., 2010).

However, there is also no 'one-size-fits-all' solution to making biofuels pro-poor. There are a range of crops that can be transformed into biofuels which need different agronomic conditions that can affect whether or not the outcome is pro-poor. Agricultural knowledge increasingly allows selection of those that can be considered pro-poor, e.g. sweet sorghum. Biofuel crops integrated into decentralised small-scale non-irrigated food and fuel smallholder production systems may be considered pro-poor. Perennial crops generally have better levels of biodiversity than annual crops, which in turn can also have benefits for the crop yields. Pro-poor outcomes are also more likely when there is horizontal integration which brings additional benefits, for example, better terms of inclusion into BVCs; also when rural people own other higher value components of the chain.

Rural dwellers probably most at risk from biofuels are the agricultural workers on large plantations who are vulnerable to exploitation. (Although it should be stressed again that this exploitation is nothing innate in biofuels but it is a characteristic of the system of production.) In this context, and also in the need to protect local level ecosystem services, is it possible that the interests of the poor become the interests of the powerful? There are some who consider that CSR can be used as an instrument to achieve those ends. CSR is generally seen as a mechanism attempting to curb specific types of malpractice and improve aspects of social performance. However, what is improved seems to be selective with a focus on health and safety, while neglecting rights of collective bargaining and representation (Prieto-Carrón *et al.*, 2006). This raises questions about who has the power to make decisions, what power structures are implicit in CSR and who has a voice in the debate. There has been the limited participation of actors from the South, particularly women and other marginalised groups, in consultation and decision-making processes associated with CSR and in the design of CSR instruments and institutions (Utting, 2007). This critique can certainly call into question the role of CSR as an instrument that ensures earlier injustices are not repeated, rather that it should become an instrument of the powerful to legitimise their actions.

Undoubtedly the resourcefulness of the rural poor should not be underestimated when it comes to inclusion in BVCs:

> There are solid grounds for regarding resource managers, even poor ones, as autonomous, responsible, experimental and, though risk averse, also opportunistic. Constraints, not ignorance, deter poor households. It follows from such an opportunistic interpretation that they don't need to be cajoled, pressured or motivated but offered choices of technology or mode, access them, information, experience and the enabling economic environment that makes the effort worthwhile.
>
> (Mike Mortimer in Frost *et al.*, 2007, cited in
> Shackleton *et al.*, 2008: 112)

Terms of inclusion can be a deciding factor in poverty reduction. However, negotiating the terms of inclusion can be problematic since these negotiations are not taking place between equals. In general, in the South, and particularly in Africa, financial capital is scarce, not land or labour. This gives the upper hand to the outside investor and the smallholder and other rural actors are left with the bargaining chip of 'social licence' (McCarthy and Zen, 2010). Also there are inequalities. Therefore, policies and interventions need to support and build the capacity of the rural poor to adapt and create and build their own opportunities, rather than impose external prescriptions or uni-dimensional opportunities. It should also be accepted that the rural poor may not want to be included in BVCs.

One of the greatest threats from biofuels arises because of the pace of change in rural areas. Investment decisions have to be made quickly about feedstock crops that might take several years for maturity with all the inherent risks to farmers this entails. These risks are greatest for smallholders without a reserve of assets to tide them over until they can sell their first crop. Quick decisions do not necessarily lead to sustainable outcomes. Negative voices about biofuel programmes might bring the benefit of slowing these programmes so that more measured decisions can be made. Decisions based on the wealth of experience with agricultural commodities and small producers, about the type of technical and institutional investments required for more equitable access to the benefits of the biofuel value chain. However, if these voices bring a moratorium on all biofuel programmes in the South, they risk the possibility that the rural poor will not benefit.

Perhaps biofuels and rural poverty is best summed up by this quote from UN-Energy:

> At their best, liquid biofuels programmes can enrich farmers by helping to add value to their products. But at the worst, biofuel programmes can result in concentration of ownership that could drive the world's poorest farmers off their land and deeper into poverty.
>
> (UN–Energy, 2007: 24)

All efforts in any biofuels programme should be directed at ensuring that the former and not the latter is the outcome. In this book we have given some indications as to how that can be achieved.

Notes

1 Introduction

1 There is some inconsistency in nomenclature. In this book biofuels are classified as a subset of bioenergy. Bioenergy is defined as the group of solid, liquid and gaseous fuels derived from biomass. Some authors use the term biofuels for all fuels derived from biomass. Some authors use the term 'agrofuels' instead of 'biofuels'. This term was first coined by social movements in Latin America to describe the liquid fuels derived from food and oil crops produced in large-scale plantation-style industrial production systems. The use of the term 'agrofuels' aims to draw attention to the further entrenchment of the agribusiness model based on industrial agriculture, which includes the use of monocultures, genetically engineered plants and agro-chemicals, on farming systems in the South. While recognising and respecting this position, we have chosen to keep with the more widely used term since we explore other modes of production systems.

2 Zimbabwe, Malawi, Kenya, Philippines and Thailand.

3 The terms 'North' and 'South', first popularised by the Brandt Report (1980), used here are not meant in a geographical sense. There is no simple, agreed, unambiguous term to describe or attempt to classify the complex reality of a group of countries, which differ in degrees over a range of characteristics, from climate to culture, and with different historical experiences. The terms 'North' and 'South' have the advantage that they are value free, as say compared to 'developed' and 'developing'. The terms should be seen as convenient shorthand while keeping in mind that the groups referred to are not homogeneous entities even in terms of energy demand, supply and use.

4 Also known as vinasse.

5 The gas produced is biogas, a mixture of methane and carbon dioxide. The conversion process, anaerobic digestion, is a mature technology.

6 The volume of stillage that can be applied and hence the resulting productivity increases depends on the soil chemistry and the groundwater level (Zandbergen, 1993).

7 The costs of the pre-treatment phase equipment can make up to between 15 and 20 per cent of the total ethanol plant cost (Clancy, 1991).

8 Agave is used to make tequila.

9 'Cellulose' in Encyclopædia Britannica. [Online] http://britannica.com/EBchecked/topic/101633/cellulose, (accessed 2 October 2009).

10 For an overview of techniques see Worldwatch Institute (2007).

11 'Biodiesel'. [Online] http://en.wikipedia.org/wiki/Biodiesel#Historical_background, (accessed 14 November 2009).

12 http://gas2.org/2008/03/26/top–15-unexpected-uses-for-biodiesel/, (accessed 14 November 2009).

13 http://auto.howstuffworks.com/fuel-efficiency/alternative-fuels/biodiesel3.htm, (accessed 15 November 2009).

14   Also known as straight vegetable oil (SVO).
15   See for example, [Online] http://elsbett.com/us/elsbett-diesel-technology/elsbett-engine.html, (accessed 16 November 2009).
16   'The Caribbean faces prospect of a life without sugar.' Richard Lapper, *Financial Times*, 14 August 2005. [Online] http://globalexchange.org/campaigns/wto/3421.html, (accessed 19 November 2009).
17   'Malawi: Turmoil as tobacco prices fluctuate.' Pilirani Semu-Banda, 16 May 2008. [Online] http://afrika.no/Detailed/16690.html, (accessed 19 November 2009).
18   See for example, Patrick Barta, 'Jatropha Plant Gains Steam in Global Race for Biofuels'. *Wall Street Journal* [online edition] http://online.wsj.com/article/SB118788662080906716.html?mod=googlenews_wsj, (accessed 18 October 2009).
19   In some places soap is made from the oil.
20   http://mg.co.za/article/2007–09–25-Jatropha-fuel-for-thought, (accessed 19 October 2009); http://25degrees.net/index.php?option=com_zine&view=article&id=325: common-misconceptions-about-Jatropha&Itemid=81, (accessed 19 October 2009).
21   http://gmo-compass.org/eng/grocery_shopping/crops/19.genetically_modified _ soybean.html, (accessed 17 October 2009).
22   http://sucrose.com/home.html, (accessed 16 November 2009).
23   Although some countries, such as South Africa and Kenya, have outgrower schemes where smallholders grow sugarcane for a central processing facility.
24   Countries covered by the agreement include Belize, Costa Rica, El Salvador, Guatemala, Guyana, Honduras, Nicaragua and Panama, and the Caribbean countries Antigua, Aruba, the Bahamas, Barbados, British Virgin Islands, Dominica, Dominican Republic, Grenada, Haiti, Jamaica, Montserrat, Netherlands Antilles, St. Kitts and Nevis, St. Lucia, St. Vincent and the Grenadines, and Trinidad and Tobago (Dufey, 2006).
25   Moreira and Goldemberg (1999) report an even earlier use of bioethanol (1903). During World War I the use of bioethanol was compulsory in some areas of the country.
26   http://indexmundi.com/commodities/?commodity=rapeseed-oil&months=120, (accessed 13 October 2009). From data compiled by the IMF. The price per metric tonne of unrefined rapeseed oil FOB Rotterdam in November 2000 was US$315, from which it showed a steady increase until August 2007 when it showed sharp increases over several months peaking in July 2008. By September 2009, the price had fallen by around 50 per cent.
27   Indeed, the government of Iceland had been taking the concept of a hydrogen economy very seriously having signed an official document proclaiming its desire to transform Iceland into a hydrogen-powered society until the 2008 financial crisis set back the transformation. 'Sinking Finances Throw Iceland's "Hydrogen-Based Economy" into the Freezer', *New York Times* [Online] http://nytimes.com/cwire/2009/07/01/01climatewire-sinking-finances-throw-icelands-hydrogen-bas–47371.html, (accessed 13 September 2009).
28   http://news.bbc.co.uk/2/hi/business/7486764.stm, (accessed 5 August 2009).
29   See for example, European Energy Policy: http://ec.europa.eu/energy/index_en.htm; US Energy policy: http://eere.energy.gov/.
30   For example, 40 per cent of US Gulf of Mexico oil production being taken off-line for 30 days.
31   http://threadneedle.co.uk/media/1775136/en_threadneedle_thinks_commodities_october_2011.pdf, (accessed 19 December 2011).
32   Particulates, also known as particle matter or PM, are a complex mixture of extremely small particles and liquid droplets made up of a number of components, including acids (such as nitrates and sulphates), organic chemicals, metals and soil or dust particles. There is considerable concern about particles of ten micrometres or smaller in diameter, since these particles are small enough to pass through the throat and nose and enter the lungs leading to serious health effects. http://epa.gov/air/particlepollution/, (accessed 28 December 2011).

33 'What is the Problem with MTBE in Gasoline?' http://auto.howstuffworks.com/fuel-efficiency/fuel-consumption/question347.htm, (accessed 23 October 2009).

34 Tropical biomass is on average five times more productive than temperate biomass (Johnson and Yamba, 2005).

35 See for example, 'Biofuels Put Bucks over Ducks – After the Midwest Floods, the Corn-ethanol Lobby Wants Conserved Land Plowed Up', *Christian Science Monitor*, 24 June 2008. [Online] http://csmonitor.com/2008/0624/p08s01-comv.html, (accessed 19 November 2009).

36 http://iea.org/textbase/work/2002/ccv/ccv1%20echeverri.pdf, (accessed 19 October 2009).

37 See for example Hall, D. O. (1982; 1984).

38 'Editorial, Biofuels: Let's Look before We Leap'. *Science and Development Network* 6 December 2007. [Online] http://scidev.net/editorials; Cotula *et al.*, 2008: 14.

39 See for example http://news.bbc.co.uk/2/hi/americas/7065061.stm, (accessed 6 August 2009).

40 Using best available agricultural census data, Johnston and his co-workers consider that yields from biofuels have been overestimated because global aggregates are used which disguise considerable regional variations (Johnston *et al.*, 2009). This would imply that the area dedicated to biofuel crops has been underestimated.

41 The Subsidiary Body on Scientific, Technical and Technological Advice, Convention on Biological Diversity, Recommendation XII/7 Biodiversity and biofuel production. [Online] http://cbd.int/recommendation/sbstta/?id=11466, (accessed 28 December 2011).

42 Brazil was considered the exception since bioethanol is already a commercial product there.

43 Child Labor in Commercial Agriculture. US Department of Labor. [Online] http://dol.gov/ilab/media/reports/iclp/sweat2/commercial.htm, (accessed 19 November 2009).

44 Palm oil in the Aguan Valley, Honduras: CDM, biodiesel and murders. Biofuelwatch. [Online] http://biofuelwatch.org.uk/2011/palm-oil-in-the-aguan-valley-honduras-cdm-biodiesel-and-murders/, (accessed 29 December 2011).

45 There is no universally agreed definition of a 'livelihood'. In this book the following definition is used: a livelihood comprises the assets (natural, physical, human, financial and social capital), the activities, and the access to these (mediated by institutions and social relations) that together determine the living gained by the individual (Ellis, 2000: 10).

46 This is adapted from Evans, 2007, cited in CPRC, 2008: 25. Evans referred only to 'policies'. We have chosen a more explicit description.

2 Energy and rural poverty

1 http://unstats.un.org/unsd/demographic/sconcerns/densurb/densurbmethods.htm#B, (accessed 21 November 2009).

2 Poverty rate is defined as the percentage of the population living below the poverty line.

3 A quintile is a statistical term. It is the name given to the points 1/5, 2/5, 3/5 and 4/5 of the way through a frequency distribution. For instance, the first quintile is the point with 1/5 of the data below it and 4/5 above it. http://thesaurus.maths.org/mmkb/entry.html?action=entryById&id=470, (accessed 22 November 2009).

4 A watershed is the area of land where all of the water that is under it or drains off from it goes into the same place, such as a stream, a lake or a river which in turn act as water sources for people and aquatic habitats for plants and animals. A number of factors affect the rate at which the water drains off, including the biomass ground cover of plants and trees. Removing the biomass can speed the release of water into streams etc., which can lead to flooding, and the water storage and aquatic habitation functions are disrupted.

5 See for example the work of UNICEF http://unicef.org.uk/publications/pdf/ ECECHILD2_A4.pdf and the ILO http://ilo.org/ipec/lang--en/index.htm, (accessed 30 November 2009).

6 Although, the World Summit on Sustainable Development in 2002 made up for the oversight by adopting a plan of action to 'improve access to reliable and affordable energy services for sustainable development sufficient to facilitate the achievement of the Millennium Development Goals, including the goal of halving the proportion of people in poverty by 2015, and as a means to generate other important services that mitigate poverty, bearing in mind that access to energy facilitates the eradication of poverty'.

7 Low income households would in all likelihood be classified as 'poor'.

8 Another energy carrier purchased by households is dry cell batteries to power radios. Barnett (2000) quotes survey data from Uganda in 1996, which showed that 94 per cent of rural households not connected to the electricity grid used dry cell batteries and these households were estimated to be spending about US$6 per household per month on batteries. Barnett comments that although such batteries are convenient, they are a very expensive way of buying electricity, in terms of energy supplied, the electricity cost at the time more than US$400 per kWh.

9 A common fear expressed by development workers is that electricity may add to the burden of a woman's working day. There are mixed findings on this. A study into the socio-economic impacts of rural electrification in Namibia showed that women did stay up later than men, not working but socialising (Wamukonya and Davis, 1999).

10 Between 1990 and 2005, the total number of motor vehicles has roughly tripled in India and has increased tenfold in China (Pucher *et al.*, 2007).

11 Chinese consumption of oil in 2007 was 870,000 barrels per day higher than it had been in 2005 (Hamilton, 2009).

3 Impacts of large-scale liquid biofuel production

1 http://foodnavigator.com/Legislation/Sugar-reforms-prove-bittersweet-for-ACP, (accessed 28 December 2009).

2 The reason for the low yield is not clear. It could be climate related. It could be the type of soy seed used. If the variety requires high chemical inputs, such as fertilisers and herbicides, the *campesinos* may not be able to afford them and reduce the level of application.

3 The authors give no data for the actual number of women entering the workforce at that time.

4 Biofuel production and ecosystem services

1 An ecosystem can be defined as a dynamic complex of plant, animal, and micro-organism communities and their non-living environment interacting as a functional unit (Millennium Ecosystem Assessment, 2003).

2 Here the generic term 'ecosystem services' is used to cover both goods and services, which is in line with common usage in much of the literature.

3 Such a conceptualisation is the basis of the Gaia hypothesis formulated in the 1970s by the chemist James Lovelock and further developed by the microbiologist Lynn Margulis. See for example http://gaiatheory.org/, (accessed 5 December 2011).

4 Indeed scientists only began to conceptualise ecosystems in the 1930s (Millennium Ecosystem Assessment, 2003).

5 There are integrative approaches which bring together social science and ecological science under one framework (for example political ecology).

6 The Millennium Ecosystem Assessment (MEA) was established by United Nations Secretary General Kofi Annan in June 2001, to provide scientific information on the links between ecosystem change and human well-being to meet the needs of decision

makers. The Parties to the Convention on Biological Diversity also use the information produced by the MEA in the implementation of their treaty.

7　Evapotranspiration (ET) is the process that combines evaporation and transpiration. Evaporation means that water is converted to water vapour and removed from the evaporating surface. Water evaporates from a variety of surfaces, such as lakes, rivers, pavements, soils and wet vegetation. Transpiration consists of the vaporisation of liquid water contained in plant tissues and the vapour removal to the atmosphere.

8　The Proálcool programme is the Brazilian Alcohol National Programme established in 1975 aimed at promoting the national production of alcohol with the purpose of attending to the needs of the national and international market and the national policy on transport fuels.

9　Deepwater Horizon oil spill. http://en.wikipedia.org/wiki/Deepwater_Horizon_oil_spill, (accessed 14 December 2011).

10　A biodiversity hotspot can be defined as an area where there are exceptional concentrations of endemic species experiencing loss of habitat (Myers *et al.*, 2000: 853).

5　Liquid biofuel production and rural communities' food security

1　The Plan of Action together with the Rome Declaration on World Food Security is the output of the World Food Summit held in Rome, November 1996. [Online] http://fao.org/docrep/003/w3613e/w3613e00.htm, (accessed 1 May 2010).

2　http://un.org/en/documents/udhr/index.shtml#a11, (accessed 1 May 2010).

3　'Chronic food insecurity occurs when people are unable to access sufficient, safe and nutritious food over long periods, such that it becomes their normal condition' (UN Millennium Project, 2005: 20).

4　Amartya Sen won the Nobel Prize in Economics in 1998 for his contribution to welfare economics and for his work on famine, human development theory, and the underlying mechanisms of poverty.

5　There are 33 countries from sub-Saharan Africa included in the UN list of least developed countries. Sub-Saharan Africa has 204 million people regarded as hungry and it is the only region of the world where hunger is increasing (UN Millennium Project, 2005).

6　World Food Programme/INTERFAIS, Quantity Reports, WFP 2006, 2007, 2008.

7　It is interesting to note that while it is quite common to 'blame' India and China in this context, FAO does not consider that, at least for cereals, these two countries are culpable. Indeed India has only once imported grains this century and China has been a net exporter. However, their importation of vegetable oils into the two countries has increased; but again, FAO consider the levels not to be a significant factor in the dramatic price increases of these commodities (FAO 2008b).

8　For example, between January 2002 and February 2008 the dollar fell around 34 per cent against the euro (Mitchell, 2008: 11).

9　The exact size of investment is difficult to give but it has been estimated that in mid-2008 it was in the range of US$250–300 billion (Baffles and Haniotis, 2010).

10　The word 'managed' is used here deliberately since in many cultures women do not own land. However, as part of the unwritten gender contract that exists in rural societies women will have rights to farm certain plots of land, usually for subsistence crops to supply the household. The exact construction of these rights varies from culture to culture and these rights can be in conflict with formal legal rights.

11　CONAB is a public owned company, attached to the Brazilian Ministry of Agriculture, Livestock and Food Supply, in charge of the execution of the public policies for agriculture and supply.

12　http://gas2.org/2008/04/14/perfect-storm-inflating-food-prices-worldwide/, (accessed 5 October 2010).

13　A perfect storm is a term used to describe an event where a rare combination of circumstances will aggravate a situation drastically (Wikipedia; accessed 5 October 2010).

14 The World Bank estimates that international remittances totalled US$420 billion in 2009, of which US$317 billion went to developing countries, involving some 192 million migrants or 3.0 per cent of world population and their families. [Online] http://remittanceprices.worldbank.org/, accessed (5 October 2010).

15 'S[outh] Africa's maize plantings seen up, wheat output down'. Reuters Africa 25 October 2011, (accessed 30 December 2011); http://.afriqueavenir.org/en/2011/02/16/successes-in-african-agriculture-the-expansion-of-maize-production/, (accessed 30 December 2011).

16 'Soaring food prices and the rural poor: feedback from the field'. IFAD (undated). [Online] http://ifad.org/operations/food/food.htm, (accessed 30 December 2011).

6 Do liquid biofuels address rural energy and poverty issues?

1 See HEDON, the household energy network, for comprehensive documentation on stoves. [Online] http://hedon.info/

2 'Poor quality' biomass would mean its combustion characteristics would be poor in terms of not matching cooking requirements, burning either too fast or too slow, and with emission levels that are injurious to health.

3 The evidence would suggest that a decline in family size is linked to women's access to information about health issues, including how to control fertility, rather than TV providing an alternative to sex! (World Bank, 2008).

4 Indeed, an early excerpt of the World Energy Outlook 2010, produced for the UN General Assembly on the Millennium Development Goals, in its chapter on Energy Poverty refers to household energy in terms of access to clean energy for cooking and electricity. [Online] http://worldenergyoutlook.org/, (accessed 26 October 2010).

5 While at first sight this may not be considered serious, since it is for a non-edible end-use, depending on the adulterant, there could be engine damage as well as exposure to combustion products that can be damaging to health.

6 This is not the norm. Utilities have been criticised for their negative attitudes to rural electrification particularly for their lack of innovation in assisting low income households to obtain a connection (Kooijman and Clancy, 2010).

7 There is no agreed definition of what constitutes a micro- and minidistillery. As a rule of thumb a microdistillery produces up to 500 litres per day of fuel grade ethanol and a minidistillery could be considered to produce up to ten times as much. Ortega *et al.* (undated) estimate that a microdistillery, in Brazil, would use around 0.5 tonnes of sugarcane grown on approximately 4ha to produce 500 litres of ethanol per day, while the microdistillery would require proportionally more resources.

8 There is no universally agreed definition of CSR, however the following definition is indicative of the concept:

> an umbrella term for a variety of theories and practices all of which recognize the following: (a) that companies have a responsibility for their impact on society and the natural environment, sometimes beyond legal compliance and the liability of individuals; (b) that companies have a responsibility for the behaviour of others with whom they do business (e.g. within supply chains); and that (c) business needs to manage its relationship with wider society, whether for reasons of commercial viability, or to add value to society .
>
> (Prieto-Carrón *et al.*, 2006)

9 The biodiesel is a mixture of 30 per cent fossil diesel and 70 per cent unrefined jatropha oil. Unrefined oil is used to reduce the complexity of operation since the primary focus of the project is to reduce the drudgery of processing shea nuts (Karlsson and Banda, 2009).

10 The ecosystem approach was adopted in 2000 by the Conference to the Parties of the United Nations Convention on Biological Diversity as the main framework for implementing the Convention on Biological Diversity.

11 Not all of the palm oil is destined for biodiesel, as there are parallel markets for cooking oil and for export to the cosmetics/pharmaceuticals markets.
12 http://viacampesina.org/en/, (accessed 31 December 2011).
13 http://practicalaction.org/, (accessed 31 December 2011).
14 Based on a historical analysis of rural electrification in industrialised countries, it was concluded that economic circumstances were not a decisive factor in wide-scale rural electrification but lobbies and pressure groups were probably a larger determinant (Zomers, 2001).

7 Can biofuels be made pro-poor?

1 Sachs I., quoted in 'An In-depth Look at Brazil's Social Seal Fuel'. [Online] http://biopact.com/2007/03/in-depth-look-at-brazils-social-fuel.html, (accessed 9 September 2007).
2 The concept of a 'value chain' refers to the sequence of activities that links the primary producer to the consumer in which value is added at each step.
3 Here 'institutions' is used in the sense of 'rules of the game' as described by North (North, 1990).
4 http://biopact.com/2007/03/in-depth-look-at-brazils-social-fuel.html, (accessed 9 September 2007)
5 The cases selected included more than biofuels. Many of the projects were still in the early stages so further lessons will probably emerge. They were specifically selected on the basis of local participation and leadership.
6 The role of the NGO and CBOs in the 14 case studies varied including contract negotiation, financing and quality control.
7 http://www.fsc.org.
8 *New Straits Times* (2008) 'Winning 'Em Over with Clean Green', 9 July.
9 At the time of writing (early 2009), the Roundtable on Sustainable Biofuels had responded to these criticisms and had widely circulated its Standard for Sustainable Biofuels. [Online] http://cgse.epfl.ch/page65660.html.
10 A *panchayat* is the lowest level of the state administration system in India.
11 The notable exception is bioethanol in Brazil.
12 The right to organise and collective bargaining are enshrined in the International Labour Organization 1949 Convention: the Application of the Principles of the Right to Organise and to Bargain Collectively. [Online] http://ilo.org/ilolex/cgi-lex/convde.pl?C098, (accessed 22 June 2012).
13 The authors do not specify 'small'.
14 See for example, the Great Famine of the 1840s in Ireland, when the majority of the poor were starving due to continued failure of the potato crop, their basic food, while wheat grown in Ireland was exported to England (Woodham-Smith, 1962).

References

Abramovay, R. and Magalhães, R. (2007) 'Access of family farmers to biodiesel markets: partnerships between large companies and social movements', *Regoverning Markets Programme*. University of São Paolo, Brazil

Achten, W. M. J., Verchot, L., Franken, Y. J., Mathijs, E., Singh, V. P., Aerts, R. and Muys, B. (2008) 'Jatropha biodiesel production and use', *Biomass and Bioenergy* 32: 1063–84

Agarwal, B. (1997) 'Gender, environment and poverty interlinks: regional variations and temporal shifts in rural India', 1971–91 *World Development* 25: 23–52

Aitken, J. M., Cromwell, G. and Wishart, G. (1991) *Mini and Micro-hydropower*. ICIMOD, Kathmandu, Nepal

Amacher, G. S., Hyde, W. F. and Joshee, B. R. (1993) 'Joint production and consumption in traditional households: fuelwood and crop residues in two districts in Nepal', *Journal of Development Studies* 30: 206–25

Amezaga, J. M., Von Maltitz, G. and Boyes, S. L. (2010) *Assessing the Sustainability of Bioenergy Projects in Developing Countries: A Framework for Policy Evaluation*. Newcastle University, Newcastle, UK

Amigun, B., Sigamoney, R. and Von Blottnitz, H. (2008) 'Commercialisation of biofuel industry in Africa: a review', *Renewable and Sustainable Energy Reviews* 12: 690–711

Amnesty International (2008) 'The state of the world's human rights: Amnesty International Report 2008'

Anderson, S. A. (1990) 'Core indicators of nutritional state for difficult to sample populations', *Journal of Nutrition* 102: 1559–660

Apple, J., Vicente, R., Yarberry, A., Lohse, N., Mills, E., Jacobson, A. and Poppendieck, D. (2010) 'Characterization of particulate matter size distributions and indoor concentrations from kerosene and diesel lamps', *Indoor Air* 20: 399–411

Ariffin, D., Idris, A. S. and Singh, G. (2000) 'Status of ganoderma in oil palm', in Flood, J., Bridge, P. D. and Holderness, M. (eds) *Ganoderma Diseases of Perennial Crops*. CABI Publishing, Wallingford, UK

Ariza-Montobbio, P., Lele, S., Kallis, G. and Martinez-Alier, J. (2010) 'The political ecology of Jatropha plantations for biodiesel in Tamil Nadu, India', *Journal of Peasant Studies* 37: 875–97

Ash, N. and Jenkins, M. (2007) *Biodiversity and Poverty Reduction: The Importance of Biodiversity for Ecosystem Services*. UNEP-WCMC, Cambridge

Ashley, C. and Maxwell, S. (2001) 'Rethinking rural development', *Development Policy Review* 19(4): 395–425

Baffes, J. and Haniotis, T. (2010) 'Placing the 2006/08 commodity price boom into perspective', *Policy Research Working Paper 5371*. The World Bank, Washington DC

Baier, S., Clements, M., Griffiths, C. and Ihrig, J. (2009) 'Biofuels impact on crop and food prices: using an interactive spreadsheet', *International Finance Discussion Papers Number 967*. Board of Governors of the Federal Reserve System, Washington DC

Bailey, R. (2008) *Another Inconvenient Truth: How Biofuel Policies are Deepening Poverty and Accelerating Climate Change*. Oxfam International, Oxford

Bakker, S. J. A. (2006) *CDM and Biofuels: Can the CDM Assist Biofuel Production and Deployment?* Energy Research Centre of the Netherlands (ECN), Petten, The Netherlands

Balch, O. and Carroll, R. (2007) 'Massacres and paramilitary land seizures behind the biofuel revolution', *Guardian* (UK): 5 June 2007

Barnes, D. and Foley, G. (2004*) Rural Electrification in the Developing World: A Summary of Lessons from Successful Programs*. UNPD, New York, and World Bank, Washington DC

Barnett, A. (1999) 'Increasing access to sustainable energy sources': A summary of recommendations made to the Sustainable Energy Programme of the Shell Foundation

Barnett, A. (2000) 'Energy and the fight against poverty', unpublished paper presented at the Institute for Social Studies, The Hague

Baxter, A. (2010) 'Socio-economic implications of biofuel Africa's Jatropha development for local communities in northern Ghana', BSc in Environmental Science Research, Aberdeen, UK

Bendz, K. (2005) 'Pakistan, EU's second largest ethanol exporter, loses privileged status', in U.F.A. Service (ed.) *GAIN Report*, Global Agriculture Information Network

Bennett, E. M., Peterson, G. D. and Levitt, E. A. (2005) 'Looking to the future of ecosystem services', *Ecosystems* 8: 125–32

Berndes, G. (2002) 'Bioenergy and water – the implications of large-scale bioenergy production for water use and supply', *Global Environmental Change* 12: 253–71

Bickel, U. and Dros, J. M. (2003) *The Impacts of Soybean Cultivation on Brazilian Ecosystems*, World Wildlife Fund. [Online] http://assets.panda.org/downloads/impactsofsoybean.pdf (accessed 1 February 2012)

Biofuels Research Advisory Council (2006) *Biofuels in the European Union: A Vision for 2030 and beyond*. European Commission, Brussels

Blowfield, M. (2007) 'Reasons to be cheerful? What we know about CSR's impact', *Third World Quarterly* 28: 683–95

Bolwig, S., Ponte, S., Du Toit, A., Riisgaard, L. and Halberg, N. (2008) 'Integrating poverty, gender and environmental concerns into value chain analysis: a conceptual framework and lessons for action research', *DIIS Working Papers 2008/16*. Copenhagen, Danish Insitute for Inernational Studies, Denmark

Borras, S. M., McMichael, P. and Scoones, I. (2010) 'The politics of biofuels, land and agrarian change: editors' introduction', *Journal of Peasant Studies* 37: 575–92

Bouis, H. and Haddad, L. J. (1994) 'The nutrition effects of sugarcane cropping in a southern Philippine province', in Von Braun, J. and Kennedy, E. (eds) *Agricultural Commercialization, Economic Development, and Nutrition*. The Johns Hopkins University Press, Baltimore, MD

Boyd, J. and Banzhaf, S. (2007) 'What are ecosystem services? The need for standardized environmental accounting units', *Ecological Economics* 63: 616–26

Broca, S. S. (2002) *Food Insecurity, Poverty and Agriculture: A Concept Paper*, ESA Working Paper No. 02–15, FAO, Rome

Burley, H. and Griffiths, H. (2009) *Jatropha: Wonder Crop? Experience from Swaziland*. Friends of the Earth, Amsterdam

Burn, N. and Coche, L. (2001) 'Multifunctional platform for village power', in Misana, S. and Karlsson, G. V. (eds) *Generating Opportunities: Case Studies on Energy and Women*. UNDP, New York

Buyx, A. and Tait, J. (2011) 'Ethical framework for biofuels', *Science* 332: 540–1

Casson, A. (2000) *The Hesitant Boom: Indonesia's Oil Palm Sub-Sector in an Era of Economic Crisis and Political Change*. Center for International Forestry Research (CIFOR), Bogor, Indonesia

Cecelski, E. W. (1995) 'From Rio to Beijing: engendering the energy debate', *Energy Policy* 23: 561–75

Cecelski, E. W. (2000) 'Energy and poverty reduction: the role of women as a target group', Debate on Sustainable Energy in Danish Development Assistance. ENERGIA, Copenhagen

CFC (2007) 'Biofuels: strategic choices for commodity dependent developing countries', *Commodities Issues Series*. Common Fund for Commodities, Amsterdam

Chege, K. (2007) *Biofuel: Africa's New Oil?* SciDev.Net. [Online] http://scidev.net/en/features/biofuel-africas-new-oil.html, (accessed 11 December 2007)

Christian Aid (2009) 'Growing pains: the possibilities and problems of biofuels', *Christian Aid Report 2009*, London. [Online] http://christianaid.org.uk/images/biofuels-report-09.pdf, (accessed 9 August 2012)

Clancy, J. S. (1991) 'Small scale production of fuel ethanol and its utilisation in stationary spark ignition engines', PhD, University of Reading

Clancy, J. S. (2002) 'Blowing the smoke out of the kitchen: gender issues in household energy', *SPARKNET Brief No1 Gender and Energy*. [Online] doc.utwente.nl/59060/ (accessed 8 August 2012)

Clancy, J. S. (2008) 'Are biofuels pro-poor? Assessing the evidence', *European Journal of Development Studies* 20(3): 416–31

Clancy, J. S. and Kooijman, A. (2006) *Enabling Access to Sustainable Energy: A Synthesis of Research Findings in Bolivia, Tanzania and Vietnam*. EASE, ETC Energy, Leusden, The Netherlands

Clancy, J. S., Skutsch, M. and Batchelor, S. (2003) 'The gender–energy–poverty nexus: finding the energy to address gender concerns in development', *Project CNTR998521*. DFID, London

Clancy, J. S., Oparaocha, S. and Roehr, U. (2006) 'Gender equity and renewable energies', in Assmann, D., Laumanns, U. and Uh, D. (eds) *A Global Review of Technologies, Policies and Markets*. Earthscan, London

Clancy, J. S., Winther, T., Matinga, M. and Oparaocha, S. (2011b) 'Gender equity in access to and benefits from modern energy and improved energy technologies', *World Development Report 2012* Background Paper. The World Bank, New York. [Online], http://norad.no/en/thematic-areas/energy/gender-in-energy (accessed 10 September 2012)

Clements, R. (2008) 'Scoping study into the impacts of bioenergy development on food security', Practical Action Consulting, Bourton-on-Dunsmore, UK

Cline, W. R. (2004) *Trade Policy and Global Poverty*. Center for Global Development and Institute for International Economics, Washington

Cloin, J., Woodruff, A. and Fürstenwerth, D. (2007) 'Liquid biofuels in Pacific island countries', in SOPAC (ed.) *SOPAC Miscellaneous Report*. SOPAC, Suva, Fiji Islands

Cohen, A. J., Anderson, H. R., Ostro, B. *et al.* (eds) *Comparative Quantification of Health Risks: Global and Regional Burden of Disease Attributable to Selected Major Risk Factors*. WHO, Geneva

Cohen, M. J., Tirado, C., Aberman, N.-L. and Thompson, B. (2008) *Impact of Climate Change and Bioenergy on Nutrition*. IFPRI and FAO, Rome

Colchester, M., Jiwan, N., Martua Sirait, A., Firdaus, A. Y., Surambo, A. and Pane, H. (2006) *Promised Land: Palm Oil and Land Acquisition in Indonesia – Implications for*

Local Communities and Indigenous Peoples. Forest Peoples Programme and Perkumpulan Sawit Watch

Collins, K. (2008) 'The role of biofuels and other factors in increasing farm and food prices: a review of recent development with a focus on feed grain markets and market prospects', Kraft Food Global

Commission for Africa (2005) *Our Common Interest.* Commission for Africa

Commission of the European Communities (2006) 'An EU strategy for biofuels', Commission of the European Communities, Brussels

Comprehensive Assessment of Water Management in Agriculture (CA) (2007) *Water for Food, Water for Life: A Comprehensive Assessment of Water Management in Agriculture.* Earthscan/International Water Management Institute, London/Colombo

Cooke, P., Köhlin, G. and Hyde, W. F. (2008) 'Fuelwood, forests and community management – evidence from household studies', *Environment and Development Economics* 13: 103–35

Cotula, L. and Vermeulen, S. (2009) 'Deal or no deal: the outlook for agricultural land investment in Africa', *International Affairs* 85: 1233–47

Cotula, L., Dyer, N. and Vermeulen, S. (2008) *Fuelling Exclusion? The Biofuels Boom and Poor People's Access to Land.* IIED, London

Couto, R. and Murren, J. (2009) 'Tapping the potential of Brazil's Proalcool movement for the household energy sector'. *Boiling Point* 56: 36–8

CPRC (2008) *The Chronic Poverty Report 2008–09: Escaping Poverty Traps.* The Chronic Poverty Research Centre (CPRC), Manchester

Cramer Commission (2007) *Testing Framework for Sustainable Biomass.* Energy Transition's Interdepartmental Programme Management (IPM), The Hague, The Netherlands

Cromwell, E., Luttrell, C., Shepherd, A. and Wiggins, S. (2005) *Poverty Reduction Strategies and the Rural Productive Sectors: Insights from Malawi, Nicaragua and Vietnam.* ODI, London

Daily, G. (1997) 'Introduction: what are ecosystem services?', in Daily, G. (ed.) *Nature's Services: Societal Dependence on Natural Ecosystems.* Island Press, Washington DC

Danielsen, F., Beukema, H., Burgess, N. D., *et al.* (2009) 'Biofuel plantations on forested lands: double jeopardy for biodiversity and climate', *Conservation Biology* 23: 348–58

Dasgupta, N. (1999) 'Energy efficiency and poverty alleviation', *DFID Energy Research Newsletter* 8

Dasgupta, P. (1993) *An Inquiry into Well-Being and Destitution.* Clarendon Press, Oxford

Datt, G. and Ravallion, M. (1998) 'Why have some Indian states done better than others at reducing rural poverty?' *Economica* 65(257): 17–38

Dauvergne, P. and Neville, K. J. (2010) 'Forests, food, and fuel in the tropics: the uneven social and ecological consequences of the emerging political economy of biofuels', *Journal of Peasant Studies* 37: 631–60

de Fraiture, C., Giordano, M. and Yongsong, L. (2008) 'Biofuels: implications for agricultural waste water use: blue impacts of green energy', *Water Policy Supplement* 1: 67–81

Dessie, G. and Erkossa, T. (2011) *Eucalyptus in East Africa: Socio-economic and Environmental Issues,* Working Paper FP46/E, FAO, Rome, Italy

DFID (2007) *Land: Better Access and Secure Rights for Poor People.* DFID, London

Dominguez-Faus, R., Powers, S., Burken, J. and Alvarez, J. P. (2009) 'The water footprint of biofuels: a drink or drive issue?', *Environmental Science & Technology* 43: 3005–10

Donald, P. F. (2004) 'Biodiversity impacts of some agricultural commodity production systems', *Conservation Biology* 18: 17–38

Doornbosch, R. and Steenblik, R. (2007) 'Biofuels: is the cure worse than the disease?' Roundtable on Sustainable Development. OECD, Paris, France

Dorward, A., Kydd, J., Morrison, J. and Urey, I. (2004) 'A policy agenda for pro-poor agricultural growth', *World Development* 32: 73–89

Dufey, A. (2006) *Biofuels Production, Trade and Sustainable Development: Emerging Issues.* IIED, London

Dufey, A. (2007) *International Trade in Biofuels: Good for Development? And Good for Environment?* IIED, London

Eckholm, E. (1975) 'The other energy crisis: firewood', *Worldwatch Paper* No. 1, Washington DC

Edwards, R., Szekeres, S., Neuwahl, F. and Mahieu, V. (2008) *Biofuels in the European Context: Facts and Uncertainties.* European Commission Joint Research Centre, Petten, Netherlands and Ispra, Italy

Eide, A. (2008) *The Right to Food and the Impact of Liquid Biofuels (Agrofuels).* FAO, Rome

Ellis, F. (2000) *Rural Livelihoods and Diversity in Developing Countries.* Oxford University Press, Oxford

EPA (1999) *Smog – Who Does It Hurt? What You Need to Know about Ozone and Your Health*, EPA Publication No. EPA-452/K-99-001 Washington DC. [Online] http://epa.gov/airnow//health/smog.pdf, (accessed 10 September 2012)

Ernsting, A. (2007) 'Agrofuels in Asia: fuelling poverty, conflict, deforestation and climate change', *Seedling.* GRAIN, Barcelona, Spain

ESMAP (1999) 'Household energy strategies for urban India; the case of Hyderabad', Report 214. The World Bank, Washington DC

European Commission (2006) *An EU Strategy for Biofuels.* European Commission, Brussels (COM/2006/0034)

Evans, A. (2009) *The Feeding of the Nine Billion. Global Food Security for the 21st Century.* Chatham House (Royal Institute of International Affairs), London

Evans, P. (2007) 'PRS synthesis paper. Background paper for The Chronic Poverty Report 2008–09', Chronic Poverty Research Centre (CPRC), Manchester, UK

Fairless, D. (2007) 'Biofuel: the little shrub that could – maybe', *Nature* 449(7163): 652–5

Falconer, J. and Arnold, J. E. M. (1991) *Household Food Security and Forestry – An Analysis of Socio-Economic Issues.* FAO, Rome

FAO (1989) *Prevention of Postharvest Food Losses Fruits, Vegetables and Root Crops: A Training Manual.* FAO, Rome

FAO (2008a) *The State of Food and Agriculture: Biofuels – Prospects, Risks and Opportunities.* FAO, Rome

FAO (2008b) 'Soaring food prices: facts, perspectives, impacts and actions required', *High-Level Conference on World Food Security: The Challenges of Climate Change and Bioenergy.* FAO, Rome

FAO (2009) *The State of Food Insecurity in the World.* FAO, Rome

FAO (2010) 'Bioenergy environmental impact analysis: analytical framework', *Environmental and Natural Resources Management Working Paper.* FAO, Rome

Fatimah, Y. A. (2011) 'Actors in transition: Jatropha initiatives in Indonesian villages', *Innovation and Sustainability Transitions in Asia.* Kuala Lumpur, Malaysia

Field, C. B., Campbell, J. E. and Lobell, D. B. (2007) 'Biomass energy: the scale of the potential resource', *Trends in Ecology and Evolution* 23(2): 65–72

Fischlin, A., Midgley, G. F., Price, J. T., *et al.* (2007) 'Ecosystems, their properties, goods, and services', in Parry, M. L., Canziani, O. F., Palutikof, J. P., van der Linden, P. J. and Hanson, C. E. (eds) *Climate Change 2007: Impacts, Adaptation and Vulnerability. Contribution of Working Group II to the Fourth Assessment Report of the Intergovernmental Panel on Climate Change.* Cambridge University Press, Cambridge

Fontana, M. and Paciello, C. (2009) 'Gender dimensions of rural and agricultural employment: Differentiated pathways out of poverty', FAO-IFAD-ILO Workshop on Gaps, Trends and Current Research in Gender Dimensions of Agricultural and Rural Employment: Differentiated Pathways out of Poverty. IFAD, Rome

Food Corporation of India (2003) 'Annual Report 2001–2', Department of Food and Public Distribution. Food Corporation of India, New Delhi

Freeman, L., Lewis, J. and Borreill-Freeman, S. (2008) 'Free, prior and informed consent: implications for sustainable forest management in the Congo Basin', *Workshop on Forest Governance and Decentralisation in Africa*: 8–11 April, Durban, South Africa

Friends of the Earth (2008) *Fuelling Destruction in Latin America: The Real Price of the Drive for Agrofuels*, Friends of the Earth International, Amsterdam 113: 48

Fromm, I. (2007) *Integrating Small-scale Producers in Agrifood Chains: The Case of the Palm Oil Industry in Honduras*, 17th Annual Food and Agribusiness Forum and Symposium. Parma, Italy

Frost, P., Campbell, B., Luckert, M., Mutamba, M., Mandondo, A. and Kozanayi, W. (2007) 'In search of improved rural livelihoods in semi-arid regions through local management of natural resources: Lessons from case studies in Zimbabwe', *World Development* 35: 1961–74

Frynas, J. G. and Wood, G. (2001) 'Oil and war in Angola', *Review of African Political Economy* 28(90): 587–606

Gaia Foundation, Biofuelwatch, The African Biodiversity Network, Salva La Selva, Watch Indonesia and Econexus (2008) *Agrofuels and the Myth of the Marginal Lands*. [Online] http://cbd.int/doc/biofuel/Econexus%20Briefing%20AgrofuelsMarginalMyth.pdf

Geldenhuys, C. J. (1999) 'Requirements for improved and sustainable use of forest biodiversity: examples of multiple forest use in South Africa'. In Poker, J., Stein, I. and Werder, U. (eds) *Proceedings Forum Biodiversity – Treasures in the World's Forests*, Alfred Toepfer Akademie, Schneverdingen, Germany, 72–82

Gerbens-Leenes, W., Hoekstra, A. Y. and van der Meer, T. H. (2009) 'The waterfootprint of bioenergy', *Proceedings of the National Academy of Science* 106: 10219–23

German Advisory Council on Global Change (2008) *Future Bioenergy and Sustainable Land Use*. Earthscan, London

Gibbs, H. K., Ruesch, A. S., Achard, F., Clayton, M. K., Holmgren, P., Ramankutty, N. and Foley, J. A. (2010) 'Tropical forests were the primary sources of new agricultural land in the 1980s and 1990s', *Proceedings of the National Academy of Sciences* 107: 16732–7

Gladwin, C. H. and Thomson, A. M. (1997) *Food or Cash Crops: Which is the Key to Food Security for African Women Farmers?* American Anthropological Association/International Association of Agricultural Economists, Washington DC/Sacramento CA

Global Bioenergy Parntership (2011) *The Global Bioenergy Partnership Sustainability Indicators for Bioenergy*. FAO, Rome

Goebertus, J. (2008) 'Palma de aceite y desplazamiento forzado en Zona Bananera: "trayectorias" entre recursos naturales y conflicto', *Colombia Internacional* 67: 152–75

Goldemberg, J. (2006) 'The ethanol program in Brazil', *Environmental Research Letters* 1: 5

Govereh, J. and Jayne, T. S. (1999) *Effects of Cash Crop Production on Food Crop Productivity in Zimbabwe: Synergies or Trade-Offs*. Michigan State University

Goyal, P. (2002) 'Food security in India', *The Hindu*, Jan 10. [Online] http://thehindu. com (accessed 10 September 2012)

Greenpeace (2008) *Burning Up Borneo*. Greenpeace, Amsterdam, The Netherlands

Guijt, I., Hinchcliffe, F., Melnek, M., Bishop, J., Eaton, D., Pimbert, M., Pretty, J. and Scoones, I. (1995) *Hidden Harvest: The Value of Wild Resources in Agricultural Systems*. IIED, London

Gundimeda, H. (2005) 'Can CPRs generate carbon credits without hurting the poor?' *Economic and Political Weekly* 40: 973–80

Haines-Young, R. and Potschin, M. (2010) 'The links between biodiversity, ecosystem services and human well-being', in Raffaelli, D. and Frid, C. (eds) *Ecosystem Ecology: A New Synthesis*. Cambridge University Press, Cambridge

Hall, D. O. (1982) 'Food versus Fuel, a World Problem?', in Strub, A., Chartier, P. and Schleser, G. (eds) *Energy from Biomass*. Applied Science Publishers, London

Hall, D. O. (1984) 'Biomass: fuel versus food, a world problem?', in Hall, D. O., Nyen, N. and Margaris, M. S. (eds) *Economics of Ecosystem Management*. Dr W. Junk Publisher, Dordrecht, The Netherlands

Hamilton, J. D. (2009) 'Causes and consequences of the oil shock of 2007–08', in Romer, D. and Wolfer, J. (eds) *Brookings Papers on Economic Activities*. Brookings Institution Press, Washington DC

Hansen-Kuhn, K. (2008) *Food, Farmers and Fuel: Balancing Global Grain and Energy Policies with Sustainable Land Use*. Action Aid International, Johannesburg

Haralambous, S., Liversage, H. and Romano, M. (2009) 'The growing demand for land – risks and opportunities for smallholder farmers', Roundtable organised during the 32nd session of IFAD's Governing Council. IFAD, Rome

Hart Energy Consulting and CABI (2010) *Land Use Change: Science and Policy Review*. CABI, Wallingford, UK

Hartemink, A. E. (2005) 'Plantation agriculture in the tropics: environmental issues', *Outlook on Agriculture* 34(1): 11–21

HDRC (2002) 'Economic and social impact evaluation study of the rural electrification program', Human Development Research Centre, Dhaka

Heady, A. and Fan, S. (2008) *Anatomy of Crisis – The Causes and Consequences of Surging Food Prices*. IFPRI, Washington DC

Hedlund, H. G. B. (1989) 'Cooperatives revisited'. In Hedlund, H. G. B. (ed.) *Seminar Proceedings Cooperatives Revisited*. Nordiska Afrikainstitutet, Uppsala

Heintz, J. (2008) 'Poverty, employment and globalisation: A gender perspective', *Poverty in Focus* 13, 12–13

Henning, R. K. (2009) *The Jatropha System: An Integrated Approach of Rural Development*. [Online] http://jatrophabiodiesel.org/drRKHeaning.php (accessed 3 December 2011)

Hetterschijt, W. (2009) *Mali Biocarburant SA: Making Core Business of Sustainability*. [Online] http://landcoalition.org/cpl-blog/wp-content/uploads/Mali_Biocarburant.pdf (accessed 20 September 2010)

Hill, J., Nelson, E., Tilman, D., Polasky, S. and Tiffany, D. (2006) 'Environmental, economic, and energetic costs and benefits of biodiesel and ethanol biofuels', *PNAS* 103: 11206–10

Hooijer A., Silvius M., Wösten, H. and Page, S. (2006) 'Peat CO_2 – assessment of CO_2 emissions from drained peatlands in SE Asia', *Delft Hydraulics Report Q3943*

Hospes, O. and Clancy, J. S. (2011) 'Unpacking the discourse of social inclusion in value chains, with a case study of the soy-biodiesel chain in Brazil', in Helmsing, A. H. J. and Vellema, S. (eds) *Value Chains, Inclusion and Endogenous Development: Contrasting Theories and Realities*. Routledge, London, New York

Hussain, I. (2004) *Pro-poor Interventions Strategies in Irrigated Agriculture in Asia*. International Water Management Institute, Colombo, Sri Lanka

ICRISAT (2007) 'Pro-poor biofuels outlook for Asia and Africa: ICRISAT's perspective', *Working Paper* 13 March, ICRISAT, Hyderabad, Andhra Pradesh, India. [Online] http://icrisat.org (accessed 29 March 2009)

IDS (2003) 'Energy, poverty, and gender: a review of the evidence and case studies in rural China', Report for The World Bank. IDS, Brighton, UK

IEA (2004) *Biofuels for Transport An International Perspective*. IEA, Paris

IEA (2006) *World Energy Outlook 2006*. IEA, Paris

IEA (2008) *Energy Technology Perspectives. Scenarios and Strategies to 2050*. IEA, Paris

IEA (2009) *World Energy Outlook 2009*. IEA, Paris

IFAD (2010) *Rural Poverty Report 2011*. IFAD, Rome

IIED (2008) *Responsible Enterprise, Foreign Direct Investment and Investment Promotion: Key Issues in Attracting Investment for Sustainable Development*. IIED, London

IISD (2007) 'First high-level biofuels seminar in Africa', *International Institute for Sustainable Development Bulletin* 9(1): 26

IMF (2008) *World Economic Outlook*. IMF, Washington DC

Irwin, F. and Ranganathan, J. (2007) *Restoring Nature's Capital: An Action Agenda to Sustain Ecosystem Services*. World Resources Institute, Washington DC

IUCN (2007) 'Gender and Bioenergy', IUCN

IUCN/DFID (no date) 'Food security and biodiversity', *Biodiversity in Development-Biodiversity Brief*

Johnson, F.X. and Yamba, F. (2005) 'Comparative advantage in the production of biofuels', *Renewable Energy Partnerships for Poverty Eradication and Sustainable Development* 10, Stockholm Environment Institute

Johnston, M., Foley, J. A., Holloway, T., Kucharik, C. and Monfreda, C. (2009) 'Resetting global expectations from agricultural biofuels', *Environment Research Letters* 4(014004) 9

Jongschaap, R. E. E., Corre, W. J., Bindraban, P. S. and Brandenburg, W. A. (2007) *Claims and Facts on Jatropha curcas L.*, Plant Research International BV, Wageningen

Jull, C., Patricia, C. R., Mosoti, V. and Vapnek, J. (2007) 'Recent trends in the law and policy of bioenergy production, promotion and use', *FAO Legal Papers* 68, September

Jung, A., Dörrenberg, P., Rauch, A. and Thöne, M. (2010) *Biofuels – At What Cost? Government Support for Ethanol and Biodiesel in the European Union – 2010 Update*. IISD, Geneva

Kameri-Mbote, P. (2006) 'Women, land rights and the environment: the Kenyan experience', *Development Policy Review* 49: 43–8

Kammen, D. M., Bailis, R. and Herzog, A. V. (2001) 'Clean energy for development and economic growth: biomass and other renewable energy options to meet energy and development needs in poor countries', UNDP Environmentally Sustainable Development Group. UNDP, New York

Karlsson, G. and Banda, K. (eds) (2009) 'Biofuels for sustainable rural development and empowerment of women: case studies from Africa and Asia', *ENERGIA*. Leusden, The Netherlands

Kartha, S. and Larson, E. D. (eds) (2000) *Bioenergy Primer: Modernising Biomass Energy for Sustainable Development*. UNDP, New York

Kartha, S., Leach, G. and Rajan, S. C. (2005) *Advancing Bioenergy for Sustainable Development: Guideline for Policymakers and Investors*. The World Bank, Washington DC

Keam, S. and McCormick, N. (2008) *Implementing Sustainable Bioenergy Production: A Compilation of Tools and Approaches*. IUCN, Gland, Switzerland

Kelkar, G. (1995) 'Gender analytical tools', *Wood Energy News*, 10

Kelkar, G. and Nathan, D. (2005) *Strategic Gender Interventions and Poverty Reduction: Principles and Practice*. IFAD and UNIFEM, New Delhi, India

Kenkel, P. and Holcomb, R. B. (2006) 'Challenges to producer ownership of ethanol and biodiesel production facilities', *Journal of Agricultural and Applied Economics* 38: 369–75

Koh, L. P. (2007) 'Potential habitat and biodiversity losses from intensified biodiesel feedstock production', *Conservation Biology* 21: 1373–5

Koh, L. P. and Ghazoul, J. (2008) 'Biofuels, biodiversity, and people: understanding the conflicts and finding opportunities', *Biological Conservation* 141: 2450–60

Kojima, M. and Johnson, T. (2005) *Potential for Biofuels for Transport in Developing Countries.* ESMAP, World Bank, Washington DC

Kojima, M., Mitchell, D. and Ward, W. (2007) *Considering Trade Policies for Liquid Biofuels.* ESMAP, The World Bank, Washington DC

Kooijman-Van Dijk, A. L. (2008) 'The power to produce: the role of energy in poverty reduction through small scale enterprises in the Indian Himalayas', PhD, University of Twente, Enschede

Kooijman-Van Dijk, A. and Clancy, J. S. (2010) 'Enabling access to sustainable energy: a synthesis of research findings in Bolivia, Tanzania and Vietnam', *Energy for Sustainable Development* 14: 14–21

Koonin, S. (2006) 'Getting serious about biofuels', *Science* 311(5760): 435

Krishna, A. (2004) 'Escaping poverty and becoming poor: who gains, who loses and why?' *World Development* 32: 121–36

Kutas, G., Lindberg, C. and Steenblik, R. (2007) *Biofuels – At What Cost? Government Support for Ethanol and Biodiesel in the European Union.* IISD, Winnipeg, Canada

Lal, R. (2006) 'Land area for establishing biofuel plantations', *Energy for Sustainable Development* 10: 67–79

Landell-Mills, N. and Porras, T. I. (2002) 'Silver bullet or fools' gold? A global review of markets for forest environmental services and their impact on the poor', *Instruments for Sustainable Private Sector Forestry Series.* IIED, London

Lapola, D. M., Schaldach, R., Alcamo, J., Bondeau, A., Koch, J., Koelking, C. and Priess, J. A. (2010) 'Indirect land-use changes can overcome carbon savings from biofuels in Brazil', *Proceedings of the National Academy of Sciences* 107: 3388–93

Larson, D. and Borrell, B. (2001) 'Sugar policy and reform', in Akiyama, T., Baffes, J., Larson, D. and Varangis, P. (eds) *Commodity Market Reforms: Lessons of Two Decades.* The World Bank, Washington, DC

Lathwell, D. J. (1990) 'Legume green manures. Principles for management based on recent research', *TropSoils Bulletin*, Soil Management Collaborative Research Support Program, North Carolina State University, Raleigh, US

Laumonier, Y., Bourgeois, R. and Pfund, J.-L. (2008) 'Accounting for the ecological dimension in participatory research and development: lessons learned from Indonesia and Madagascar', *Ecology and Society* 13: 22

Lee, C. and Lazarus, M. (2011) *Bioenergy Projects and Sustainable Development: Which Project Types Offer the Greatest Benefits?* Stockholm Environment Institute, Stockholm

Legge, T. (2008) *The Potential Contribution of Biofuels to Sustainable Development and a Low-carbon Future.* Chatham House, London

Leturque, H. and Wiggins, S. (2009) *Biofuels: Could the South Benefit? The Risks Linked to Northern Targets for Biofuels Should not Overshadow Southern Opportunities.* ODI, London

Lewandowski, I. and Faaij, A. (2006) 'Steps towards the development of a certification system for sustainable bio-energy trade', *Biomass and Bioenergy* 30: 83–104

Lipsky, J. (2008) 'Commodity prices and global inflation', remarks by the First Deputy Managing Director of the IMF, the Council on Foreign Relations, New York City

Lloyd, A. C. and Cackette, T. A. (2001) 'Diesel engines: environmental impact and control', *Journal of Air and Waste Management Association* 51(6): 809–47

Lloyd, P. J. D. and Visagie, E. M. (2007) 'A comparison of gel fuels with alternative cooking fuels', *Journal of Energy in Southern Africa* 18: 26–31

López, R. (2007) 'Agricultural growth and poverty reduction'. In Bresciani, F. and Valdés, A. (eds) *Beyond Food Production: The Role of Agriculture in Poverty Reduction*. Edward Elgar Publishing, Cheltenham, UK

Lovett, J. C., Hards, S., Clancy, J. and Snell, C. (2011) 'Multiple objectives in biofuels sustainability policy', *Energy and Environmental Science* 4: 261–8

Madzwamuse, M., Schuster, B. and Nherera, B. (2007) 'The real jewels of the Kalahari. Dryland ecosystem goods and services in Kgalagadi South District, Botswana', *IUCN*. Johannesburg

Mali Biocarburant SA. (2009) *Sustainable Production of Biodiesel in West Africa*. [Online] http://malibiocarburant.com/Sustainable_production.html (accessed 21 September 2010)

Malmberg Calvo, C. (1994) 'Case study on the role of women in rural transport: access of women to domestic facilities', *Sub-Saharan Africa Transport Policy Program*. World Bank, Washington DC

Martins de Carvalho, H. (2007) *Impactos Econômicos, Sociais e Ambientais Devido à Expansão da Oferta do Etanol No Brasil*, Acción Tierra

Massé, R. and Samaranayake, M. R. (2003) 'EnPoGen study in Sri Lanka', *ENERGIA News* 5(3): 14–16

Matinga, M. N. (2010) 'We grow up with it: an ethnographic study of the experiences, perceptions and responses to the health impacts of energy acquisition and use in rural South Africa', PhD, University of Twente, Enschede

McCarthy, J. and Zen, Z. (2010) 'Regulating the oil palm boom: assessing the effectiveness of environmental governance approaches to agro-industrial pollution in Indonesia', *Law & Policy* 32: 153–79

McCornick, P. G., Awulachew, S. B. and Abebe, M. (2008) 'Water-food-energy-environment synergies and tradeoffs: major issues and case studies', *Water Policy* 10 (Suppl 1): 23–36

McMichael, P. (2010) 'Agrofuels in the food regime', *Journal of Peasant Studies* 37: 609–29

Memedovic, O. and Shepherd, A. (2009) 'Agri-food value chains and poverty reduction: overview of main issues, trends and experiences'. *Working Paper 12/2008*. Research and Statistics Branch, UNIDO, Vienna

Mertz, O., Ravnborg, H., Lövei, G., Nielsen, I. and Konijnendijk, C. (2007) 'Ecosystem services and biodiversity in developing countries', *Biodiversity and Conservation* 16: 2729–37

Millennium Ecosystem Assessment (MEA) (2003) *Ecosystems and Human Well-being: A Framework for Assessment*. Millennium Ecosystem Assessment, Washington DC

Mingorance, F. (2006) *The Flow of Palm Oil Colombia – Belgium/Europe: A Study from a Human Rights Perspective*. HREV/CBC, Brussels p. 84

Mitchell, D. (2008) 'A note on rising food prices', Development Prospects Group, *World Bank Policy Research Working Paper* 4682. World Bank, Washington DC

Molden, D., Frenken, K., Barker, R. and de Fraiture, C. (2007) 'Trends in water and agricultural development', *Water for Food, Water for Life: A Comprehensive Assessment of Water Management in Agriculture*. Earthscan, London, and International Water Management Institute, Colombo

Monsalve, S., Millán Echeverría, D. C., Flórez López, J. A., *et al.* (2008) *Agrocombustibles y Derecho a la Alimentación en América Latina. Realidad y Amenazas*. Transnational Institute – FIAN International, Amsterdam

Moreira, J. R. (2006) 'Brazil's experience with bioenergy', in Hazell, P. and Pachauri, R. K. (eds) *Bioenergy and Agriculture*. International Food Policy Research Institute, Washington DC

Moreira, J. R. (2007) 'Water use and impacts due to ethanol production in Brazil', Paper presented at International Conference on Linkages in Energy and Water Use in Agriculture in Developing Countries. IWMI and FAO, ICRISAT, Hyderabad, India

Moreira, J. R. and Goldemberg, J. (1999) 'The alcohol programme', *Energy Policy* 27: 229–45

Morris, D. (2006) 'Biofuels: what's in it for farmers and rural America?' *Oklahoma Governor's Conference on Biofuels*, University of Oklahoma

Moser, C. O. N. (1993) *Gender Planning and Development: Theory, Practice and Training.* Routledge, London

Mulat, D. (2008) 'Assessing bioethanol blending programme in Ethiopia', MSc in Environmental and Energy Management Research, University of Twente, Enschede

Mulugetta, Y., Doig, A., Dunnett, S., Jackson, T., Khennas, S. and Rai, K. (2005) *Energy for Rural Livelihoods: A Framework for Sustainable Decision Making.* ITDG Publishing, Bourton-on-Dunsmore, UK

Murren, J. and Debebe, M. (2006) 'Project Gaia's ethanol-fueled cleancook stove initiative and its impact on traditional cooking fuels used in Addis Ababa', Ethiopia, Gaia Association. [Online] http://hedon.info/ProjectGaia (accessed 12 April 2009)

Murwira, K., Wedgewood, H., Watson, C., Win, E. J. and Tawney, C. (2000) *Beating Hunger. The Chivi Experience. A Community-based Approach to Food Security in Zimbabwe.* IT Publications, London

Musafer, N. (2010) 'Community based biodiesel processing in Sri Lanka', *Boiling Point* 56, Supplementary Paper. [Online] http://hedon.info/docs/BP56_SuppPaper_PA_CommunityBiodieselSriLanka.pdf (accessed 14 April 2010)

Mwaikusa, J. T. (1993) 'Community rights and land use policies in Tanzania: the case of pastoral communities', *Journal of African Law* 37: 144–63

Myers, N., Mittermeier, R. A., Mittermeier, C. G., Da Fonseca, G. a. B. and Kent, J. (2000) 'Biodiversity hotspots for conservation priorities', *Nature* 403: 853–8

Narayan, D. (1999) *Can Anyone Hear Us? Voices from 47 Countries.* World Bank, Washington DC

Narayanaswamy, A. (2009) 'Biodiesel as an alternative fuel to petroleum diesel in Hassan', MSc Sustainable Energy Technology, University of Twente, Enschede

Narayanaswamy, A., Gowda, B. and Clancy, J. (2009) 'Biodiesel – A boon or a curse for the women of Hassan district, India?' *ENERGIA News* 12: 24–7

Nathan, D. and Nathan, G. K. (1997) 'Wood energy: the role of women's unvalued labor', *Gender Technology and Development* 1: 205–24

National Academy of Sciences (2008) *Water Implications of Biofuels Production in the United State.* The National Academies Press, Washington DC

Nelson, G. C., Rosegrant, M. W., Koo, J., Robertson, R., Sulser, T., Zhu, T., Ringler, C., Msangi, S., Palazzo, A., Batka, M., Magalhaes, M., Valmonte-Santos, R., Ewing, M. and Lee, D. (2009) *Climate Change: Impact on Agriculture and Costs of Adaptation.* IFPRI, Washington DC

Network for Social Justice and Human Rights (2007) *Agroenergy: Myths and Impacts in Latin America.* Network for Social Justice and Human Rights and Pastoral Land Commission, São Paulo

Neumann, R. P. (1995) 'Local challenges to global agendas: conservation, economic liberalization and the pastoralists' rights movement in Tanzania', *Antipode* 27: 363–82

Newell, P. and Frynas, J. G. (2007) 'Beyond CSR? Business, poverty and social justice: an introduction', *Third World Quarterly* 28: 669–81

North, D. C. (1990) *The Political Economy of Institutions and Decisions.* Cambridge University Press, Cambridge

Novo, A., Jansen, K., Slingerland, M. and Giller, K. (2010) 'Biofuel, dairy production and beef in Brazil: competing claims on land use in São Paulo State', *Journal of Peasant Studies* 37: 769–92

OECD/FAO (2007) *Agricultural Outlook 2007–2016.* OECD, Paris

OECD/IEA (2010) *Energy poverty: How to make modern energy access universal?* OECD/IEA, Paris

Oerke, E. C. and Dehne, H. W. (1997) 'Global crop production and the efficacy of crop protection – current situation and future trends', *European Journal of Plant Pathology* 103: 203–15

Ortega, E., Watanabe, M. and Cavalett, O. (undated) *Production of Ethanol in Micro and Mini-distilleries.* [Online] http://unicamp.br/fea/ortega/MarcelloMello/MicroDistillery-Ecounit.pdf, (accessed 15 April 2009)

Ortiz, R., Crouch, J. H., Iwanaga, M. *et al.* (2006) 'Bioenergy and agricultural research for development', in Hazell, P. and Pachauri, R. K. (eds) *Bioenergy and Agriculture: Promises and Challenges.* International Food Policy Research Institute, Washington DC

Oxfam (2007) *Biofuelling Poverty: Why the EU Renewable Fuel Target May be Disastrous for Poor People,* Oxfam International, Oxford

Paarlberg, R. (2010) *Food Politics: What Everyone Needs to Know.* Oxford University Press, New York

Palau, T., Cabello, D., Maeyens, A., Rulli, J. and Segovia, D. (2008) 'The refugees of the agroexport model: impacts of soy monoculture in Paraguayan campesino communities'. BASE Investigaciones Sociales

Pearce, J. (2007) 'Oil and armed conflict in Casanare, Colombia: complex contexts and contingent moments', in Kaldor, M., Karl, T. L. and Said, Y. (eds) *Oil Wars.* Pluto Press, London

Peskett, L., Slater, R., Stevens, C. and Dufey, A. (2007) 'Biofuels, agriculture and poverty reduction', *Programme of Advisory Support Services for Rural Livelihoods Final Report.* DFID, London

Pfuderer, S., Davies, G. and Mitchell, I. (2010) 'The role of demand for biofuel in the agricultural commodity price spikes of 2007/08'. DEFRA, London. [Online] http://archive.defra.gov.uk/foodfarm/food/security/price.htm (accessed 25 August 2012)

Philips, T. (2007) 'Brazil's ethanol slaves: 200,000 migrant sugar cutters who prop up renewable energy boom', *Guardian,* 9 March

Pimental, D. (2003) 'Ethanol fuels: energy balance, economics, and environmental impacts are negative', *Natural Resources Research* 12: 127–35

Pimental, D. and Patzek, T. (2005) 'Ethanol production using corn, switchgrass, and wood; biodiesel production using soybean and sunflower', *Natural Resources Research* 14: 65–76

Practical Action (2006) *Diesel Engines.* [Online] http://practicalaction.org/diesel-engines (accessed 26 October 2010)

Practical Action Consulting (2009) 'Small-scale bioenergy initiatives: brief description and preliminary lessons on livelihood impacts from case studies in Asia, Latin America and Africa'. FAO, Rome and Policy Innovation Systems for Clean Energy Security (PISCES), Nairobi

Prieto-Carrón, M., Lund-Thomsen, P., Chan, N., Muro, A. and Bhushan, C. (2006) 'Critical perspectives on CSR and development: what we know, what we don't know, and what we need to know', *International Affairs* 82: 977–87

Prowse, M. and Chihowu, A. (2007) 'Making agriculture work for the poor', *Natural Resource Perspectives.* ODI, London

Pucher, J., Peng, Z.-R., Mittal, N., Zhu, Y. and Korattyswaroopam, N. (2007) 'Urban transport trends and policies in China and India: impacts of rapid economic growth', *Transport Reviews* 27: 379–410

Rahman Osmani, S. (2010) *Food Security, Poverty and Women: Lessons from Rural Asia.* IFAD, [Online] http://ifad.org/gender/thematic/rural/rural_2.htm (accessed 5 October 2010)

Rajagopal, D. (2008) 'Implications of India's biofuel policies for food, water and the poor', *Water Policy* 10 (Suppl 1): 95–106

Rajagopal, D. and Zilberman, D. (2007) *Review of Environmental, Economic and Policy Aspects of Biofuels.* The World Bank, Washington DC

Rajvanshi, A. K. (2006) *Ethanol Fuel for Rural Households.* Boloji Media Inc. [Online] http://boloji.com/opinion/0188.htm (accessed 13 April 2009)

Ramani, K. and Heijndermans, E. (2003) *Energy, Poverty, and Gender.* The World Bank, Washington DC

Raswant, V., Hart, N. and Romano, M. (2008) 'Biofuel expansion: challenges, risks and opportunities for rural poor people', 31st session of IFAD's Governing Council, 13 February

Rathmann, R., Szklo, A. and Schaeffer, R. (2010) 'Land use competition for production of food and liquid biofuels: an analysis of the arguments in the current debate', *Renewable Energy* 35: 14–22

Ravallion, M. (2004) 'Pro-poor growth: a primer', *World Bank Policy Research Working Paper* No. 3242. World Bank, Washington DC

Reardon, T. and Barrett, C. B. (2000) 'Agroindustrialisation, globalisation, and international development: an overview of issues, patterns and determinants', *Agricultural Economics* 23: 195–205

Reardon, T. and Vosti, S. A. (1995) 'Links between rural poverty and the environment in developing countries: asset categories and investment poverty', *World Development* 23: 1495–506

Reddy, A. K. N. (2000) 'Energy and social issues', in *World Energy Assessment.* UNDP, New York

Reddy, A. K. N., Williams, R. H. and Johansson, T. B. (1997) *Energy after Rio: Prospects and Challenges.* UNDP, New York

Reddy, B. V., Rao, P. S., Kumar, A. A., Reddy, P. S., Rao, P. P., Sharma, K. K., Blummel, M. and Reddy, C. R. (2008) *Sweet Sorghum as a Biofuel Crop: Where Are We Now?* ICRISAT, Patancheru, Andhra Pradesh

Renner, M. and McKeown, A. (2010) *Promise and Pitfalls of Biofuel Jobs.* Biofuels Future Science Group

Rigg, J. (2006) 'Land, farming, livelihoods, and poverty: rethinking the links in the rural South', *World Development* 34: 180–202

Righelato, R. and Spracklen, D. V. (2007) 'Carbon mitigation by biofuels or by saving and restoring forests?' *Science* 317: 902

Rocheleau, D., Thomas-Slayter, B. and Wangari, E. (1996) 'Feminist political ecology' in Momsen, J. and Monk, J. (eds) *International Studies of Women and Place.* Routledge, London, New York

Rosegrant, M. W., Zhu, T., Msangi, S. and Sulser, T. (2008) *The Impact of Biofuel Production on World Cereal Prices.* International Food Policy Research Institute, Washington, DC

Ross, M. L. (2008) 'Blood barrels', *Foreign Affairs* 87: 2–9

Rossi, A. and Lambrou, Y. (2008) *Gender and Equity Issues in Liquid Biofuels Production: Minimizing the Risks to Maximize the Opportunities.* FAO, Rome

Rossini, R. E. and Alves Calió, S. (undated) 'Women, migration, environment and rural development policy in Brazil'. [Online] http://fao.org/docrep/x0210e/x0210e00.htm, (accessed 13 January 2008)

Royal Society (2008) *Sustainable Biofuels: Prospects and Challenges*. The Royal Society, London

RSPO (2006) 'RSPO principles and criteria for sustainable palm oil production', *Roundtable on Sustainable Palm Oil*. Selangor, Malaysia

SANDEE (2007) *Why Helping the Environment Helps Women: Understanding the Links between Resource Availability and Gender Equality in India*. South Asian Network for Development and Environmental Economics (SANDEE)

Schare, S. and Smith, K. R. (1995) 'Particulate emission rates of simple kerosene lamps', *Energy for Sustainable Development* 2: 32–6

Schnoor, J. L., Doering III, O. C., Entekhabi, D., Hiler, E. A., Hullar, T. L., Tilman, G. D. (2008) 'Environmental and economic assessment of ethanol production systems in Minnesota', Final Report to the Minnesota Pollution Control Agency. [Online] http://pca.state.mn.us/index.php/view-document.html?gid=9243, (accessed 10 September 2012)

Searchinger, T. (2009) *Evaluating Biofuels. The Consequences of Using Land to Make Fuel*. The German Marshall Fund of the United States, Washington DC

Secretaría de Agricultura y Pesca del Valle del Cauca (2008a) 'Plan de desarrollo sectorial agricola, pecuario, forestal acuicola y pesquero 2008–11', Departamento Del Valle Del Cauca, Columbia

Secretaría de Agricultura y Pesca Del Valle Del Cauca (2008b) 'Plan de inversion de la estampilla proseguridad alimentaria y desarrollo rural del Valle del Cauca 2009–11', Departamento Del Valle Del Cauca, Colombia

Sen, A. (2001) *Development as Freedom*. Oxford University Press, Oxford

Sen, G. (2008) 'Poverty as a gendered experience: the policy implications', *Poverty in Focus* 13: 6–7

Sexton, S., Zilberman, D., Rajagopal, D. and Hochman, G. (2009) 'The role of biotechnology in a sustainable biofuel future', *AgBioForum* 12: 130–40

Shackleton, C., Shackleton, S., Gambiza, J., Nel, E., Rowntree, K. and Urquhart, P. (2008) 'Links between ecosystem services and poverty alleviation: situation analysis for arid and semi-arid lands in southern Africa.' Consortium on Ecosystems and Poverty in Sub-Saharan Africa (CEPSA), Ecosystem Services and Poverty Reduction Research Programme: DIFD, NERC, ESRC, London

Shapouri, H., Duffield, J. A. and Wang, M. (2002) 'The energy balance of corn ethanol: an update', *Agricultural Report No 813*. United States Department of Agriculture, Washington DC

Sheil, D., Casson, A., Meijaard, E., Van Nordwijk, M., Gaskell, J., Sunderland-Groves, J., Wertz, K. and Kanninen, M. (2009) *The Impacts and Opportunities of Oil Palm in Southeast Asia: What do we Know and What do we Need to Know?* CIFOR, Bogor, Indonesia

Sistema Nacional de Competitividad (2009) *Produccion de Biocombustibles*. [Online] http://snc.gov.co/prensa/noticias-snc/2009/junio/nsnc_090608d.asp, (accessed 8 June 2009)

Smith, K. R. (1999) 'Indoor air pollution', *Pollution Management in Focus 4*. The World Bank, Washington DC

Snapp, S. S., Blackie, M. J., Gilbert, R. A., Bezner-Kerr, R. and Kanyama-Phiri, G. Y. (2010) 'Biodiversity can support a greener revolution in Africa', *Proceedings of the National Academy of Sciences* 107: 20840–5

Sologuren, J. (2006) 'The role of micro-finance in the up-take of electricity: A case study of small and micro-finance enterprises in rural areas of Bolivia', MSc Environmental Business Administration in Environmental and Energy Management, University of Twente, Enschede

Sparknet (2002) *South Africa Country Report Synthesis*. [Online], http://hedon.info/
    SouthAfricaCountrySynthesis (accessed 15 June 2009)

Starbuck, J. and Harper, G. D. J. (2009) *Run Your Diesel Vehicle on Biofuels: A Do-It-Yourself
    Manual*. McGraw-Hill Professional, New York

Stromberg, P. M., Gasparatos, A., Lee, J. S. H., Garcia-Ulloa, J., Koh, L. P. and Takeuchi,
    K. (2010) *Impacts of Liquid Biofuels on Ecosystem Services and Biodiversity*. United Nations
    University Institute of Advanced Studies, Yokohama, Japan

Sulle, E. and Nelson, F. (2009) *Biofuels, Land Access and Rural Livelihoods in Tanzania*.
    IIED, London

Swenson, D. (2006) 'Input-outrageous: the economic impacts of modern biofuels produc-
    tion', *Regional Science Association and the Biennial IMPLAN National Users Conference*.
    Indianapolis, IN

Swiderska, K., Roe, D., Siegele, L. and Grieg-Gran, M. (2008) 'The governance of
    nature and the nature of governance: policy that works for biodiversity and liveli-
    hoods', *Biodiversity and Livelihoods Issue Papers*. IIED, London

Swinton, S. M., Lupi, F., Robertson, G. P. and Hamilton, S. K. (2007) 'Ecosystem
    services and agriculture: cultivating agricultural ecosystems for diverse benefits',
    Ecological Economics 64: 245–52

Takavarasha, T., Uppal, J. and Hongo, H. (2005) *Feasibility Study for the Production and Use
    of Biofuel in the SADC Region*, SADC

TEEB (2010) *The Economics of Ecosystems and Biodiversity: Mainstreaming the Economics of Nature:
    A Synthesis of the Approach, Conclusions and Recommendations of Teeb*. Earthscan, London

TERI (2004) *Liquid Biofuels for Transportation: India Country Study on Potential and
    Implications for Sustainable Agriculture and Energy*. GTZ, Germany

Tewari, P. K., Batra, V. S. and Balakrishnan, M. (2007) 'Water management initiatives
    in sugarcane molasses based distilleries in India', *Resources, Conservation and Recycling*
    52: 351–67

Thomas, C. D., Cameron, A., Green, R. E., *et al.* (2004) 'Extinction risk from climate
    change', *Nature* 427: 145–8

Timmer, C. P. (2008) 'Causes of high food prices', *ADB Economics Working Paper Series*.
    Asian Development Bank, Metro Manila, Philippines

Timoney, K. P. and Lee, P. (2009) 'Does the Alberta tar sands industry pollute? The
    scientific evidence', *Open Conservation Biology Journal* 3: 65–81

Tobin, R. J. and Knausenberger, W. I. (1998) 'Dilemmas of development: burley tobacco,
    the environment and economic growth in Malawi', *Journal of Southern African Studies*
    24: 405–24

Trostle, R. (2008) *Global Agricultural Supply and Demand: Factors Contributing to the
    Recent Increase in Food Commodity Prices*. United States Department of Agriculture,
    Washington DC

Twardowska, I., Allen, H. E., Kettrup, A. A. and Lacy, W. J. (2004) *Solid Waste:
    Assessment, Monitoring and Remediation*. Elsevier BV, Amsterdam

UN Millennium Project Task Force on Hunger (2005) *Halving Hunger: It Can Be Done*.
    Earthscan, London

UNCBD (1973) *United Nations Convention on Biological Diversity*. Montreal, Canada

UNCTAD (2008) *Making Certification Work for Sustainable Development: The Case of
    Biofuels*. United Nations, New York

UNDP (1995) *Energy as an Instrument for Socioeconomic Development*. UNDP, New York

UNDP (2006) *Expanding Access to Modern Energy Services: Replicating, Scaling Up and
    Mainstreaming at the Local Level*. UNDP, New York

UN-Energy (2007) *Sustainable Energy: A Framework for Decision Makers*. UNDP, New York

UNEP (2008) 'The potential impacts of biofuels on biodiversity', *Conference of the Parties to the Convention on Biological Diversity*. UNEP, Bonn

UNEP (2009) *Towards Sustainable Production and Use of Resources: Assessing Biofuels*. UNEP Division of Technology, Industry and Economics, Paris

UNF (2008) *Sustainable Bioenergy Development in UEMOA Member Countries*. Report from United Nations Foundation, Washington DC

United Nations (2008) 'World urbanization prospects', *2007 Revision Population Database*. UN, New York

United Nations (2009) *The State of the World's Indigenous Peoples*. United Nations Department of Economic and Social Affairs, New York

University of Reading (1999) *Community Micro-Hydro in LDCs: Adoption, Management and Poverty Impact*. DFID, London

USEPA (1994) *Air Toxics from Motor Vehicles*. Office of Transportation and Air Quality, US Environmental Protection Agency, Ann Arbor, MI

Utting, P. (2007) 'CSR and equality', *Third World Quarterly* 28: 697–712

Vaidyanathan, G. (2009) 'Energizing sustainable livelihoods: a study of village level biodiesel development in Orissa, India', PhD field research. University of Waterloo, Canada

Van Gelder, J. W. and Dros, J. M. (2006) 'From rainforest to chicken breast: effects of soybean cultivation for animal feed on people and nature in the Amazon region', study commissioned by Milieudefensie / Friends of the Earth, Netherlands and Cordaid

Varghese, S. (2007) *Biofuels and Global Water Challenges*. Institute for Agriculture and Trade Policy, Minneapolis, MN

Verdonk, M., Dieperink, C. and Faaij, A. P. C. (2007) 'Governance of the emerging bio-energy markets', *Energy Policy* 35: 3909–24

Vermeulen, S. and Cotula, L. (2010) 'Over the heads of local people: consultation, consent, and recompense in large-scale land deals for biofuels projects in Africa', *Journal of Peasant Studies* 37: 899–916

Vermeulen, S. and Goad, N. (2006) *Towards Better Practice in Smallholder Palm Oil Production*. IIED, London

Virtanen, P. (2002) 'The role of customary institutions in the conservation of biodiversity: sacred forests in Mozambique', *Environmental Values* 11: 227–41

Vogt, K. A. (2007) *Forests and Society: Sustainability and Life Cycles of Forests in Human Landscapes*. CABI, Wallingford, UK

Von Braun, J. and Meinzen-Dick, R. S. (2009) *'Land Grabbing' by Foreign Investors in Developing Countries: Risks and Opportunities*. International Food Policy Research Institute, Washington DC

Walter, A., Dolzan, P., Quilodrán, O., Garcia, J., Da Silva, C., Piacente, F. and Segerstedt, A. (2008) *A Sustainability Analysis of the Brazilian Ethanol*. University of Campinas, Brazil

Wamukonya, L. and Davis, M. (1999) 'Comparisons between grid, solar and unelectrified households', *Socio-Economic Impacts of Rural Electrification in Namibia*. EDRC, University of Cape Town, South Africa,

Wandschneider, T. S. and Garrido-Mirapeix, J. (1999) *Cash Cropping in Mozambique: Evolution and Prospects*. Food Security Unit Mozambique, European Commission

Wang, M. (2005) 'The debate on energy and greenhouse gas emissions impacts of fuel ethanol', Presentation given at the Energy Systems Division Seminar. Argonne National Laboratory, Illinois, IL, August 2005

Wanitzek, U. and Sippel, H. (1998) 'Land rights in conservation areas in Tanzania', *GeoJournal* 46: 113–28

WHO (2005) *Indoor Air Pollution and Health*. World Health Organization. [Online] http://who.int/mediacentre/factsheets/fs292/en/index.html (accessed 10 March 2010)

Wickramasinghe, A. (2001) Gendered Sights and Health Issues in the Paradigm of Biofuel in Sri Lanka, *ENERGIA News* 4: 12–15

Wiggins, S., Keats, S. and Vigneri, M. (2009) *Impact of the Global Financial and Economic Situation on Agricultural Markets and Food Security*. ODI, London

Wilkinson, J. and Herrera, S. (2008) *Agrofuels in Brazil. What is the Outlook for its Farming Sector?* Oxfam International, Rio de Janeiro

Will, M. (2008) 'Promoting value chains of neglected and underutilized species for pro-poor growth and biodiversity conservation: guidelines and good practices', Global Facilitation Unit for Underutilized Species, Rome

Woodham-Smith, C. (1962) *The Great Hunger*. Penguin Books, London

World Bank (1996) 'Rural energy and development: improving energy supplies for two billion people', *Development in Practice series*. The World Bank, Washington DC

World Bank (2007) *Agriculture for Development*. The World Bank, Washington DC

World Bank (2008a) *Poverty and the Environment: Understanding Linkages at the Household Level*. The World Bank, Washington DC

World Bank (2008b) *Double Jeopardy: Responding to High Food and Fuel Prices*. The World Bank, Washington DC

World Bank (2008c) *The Welfare Impact of Rural Electrification: A Reassessment of the Costs and Benefits*. The World Bank, New York

World Food Programme (2009) 'World Food Programme Annual Report 2009', WFP, Rome

Worldwatch Institute (2007) *Biofuels for Transport*. Earthscan, London

WRI, UNDP, UNEP and World Bank (2005): 'The wealth of the poor: managing ecosystems to fight poverty', *World Resources 2005*. World Resources Institute, Washington DC

Ylonen, A. (2005) 'Grievances and the roots of insurgencies: Southern Sudan and Darfur', *Peace, Conflict and Development* 7: 40–7

Zandbergen, P. (1993) 'Energy and environmental policy in Latin America: the case of fuel ethanol in Argentina and Brazil', MSc, University of Twente, Enschede

Zhou, A. and Thomson, E. (2009) 'The development of biofuels in Asia', *Applied Energy* 86: S11–S20

Zomers, A. N. (2001) 'Rural electrification', PhD thesis, University of Twente, Enschede

Index